Other Titles in This Series

172 **K. Nomizu, Editor,** Selected Papers on Number Theory and Algebraic Geometry
171 **L. A. Bunimovich, B. M. Gurevich, and Ya. B. Pesin, Editors,** Sinai's Moscow Seminar on Dynamical Systems
170 **S. P. Novikov, Editor,** Topics in Topology and Mathematical Physics
169 **S. G. Gindikin and E. B. Vinberg, Editors,** Lie Groups and Lie Algebras: E. B. Dynkin's Seminar
168 **V. V. Kozlov, Editor,** Dynamical Systems in Classical Mechanics
167 **V. V. Lychagin, Editor,** The Interplay between Differential Geometry and Differential Equations
166 **O. A. Ladyzhenskaya, Editor,** Proceedings of the St. Petersburg Mathematical Society, Volume III
165 **Yu. Ilyashenko and S. Yakovenko, Editors,** Concerning the Hilbert 16th Problem
164 **N. N. Uraltseva, Editor,** Nonlinear Evolution Equations
163 **L. A. Bokut', M. Hazewinkel, and Yu. G. Reshetnyak, Editors,** Third Siberian School "Algebra and Analysis"
162 **S. G. Gindikin, Editor,** Applied Problems of Radon Transform
161 **Katsumi Nomizu, Editor,** Selected Papers on Analysis, Probability, and Statistics
160 **K. Nomizu, Editor,** Selected Papers on Number Theory, Algebraic Geometry, and Differential Geometry
159 **O. A. Ladyzhenskaya, Editor,** Proceedings of the St. Petersburg Mathematical Society, Volume II
158 **A. K. Kelmans, Editor,** Selected Topics in Discrete Mathematics: Proceedings of the Moscow Discrete Mathematics Seminar, 1972–1990
157 **M. Sh. Birman, Editor,** Wave Propagation. Scattering Theory
156 **V. N. Gerasimov, N. G. Nesterenko, and A. I. Valitskas,** Three Papers on Algebras and Their Representations
155 **O. A. Ladyzhenskaya and A. M. Vershik, Editors,** Proceedings of the St. Petersburg Mathematical Society, Volume I
154 **V. A. Artamonov et al.,** Selected Papers in K-Theory
153 **S. G. Gindikin, Editor,** Singularity Theory and Some Problems of Functional Analysis
152 **H. Draškovičová et al.,** Ordered Sets and Lattices II
151 **I. A. Aleksandrov, L. A. Bokut', and Yu. G. Reshetnyak, Editors,** Second Siberian Winter School "Algebra and Analysis"
150 **S. G. Gindikin, Editor,** Spectral Theory of Operators
149 **V. S. Afraĭmovich et al.,** Thirteen Papers in Algebra, Functional Analysis, Topology, and Probability, Translated from the Russian
148 **A. D. Aleksandrov, O. V. Belegradek, L. A. Bokut', and Yu. L. Ershov, Editors,** First Siberian Winter School "Algebra and Analysis"
147 **I. G. Bashmakova et al.,** Nine Papers from the International Congress of Mathematicians, 1986
146 **L. A. Aĭzenberg et al.,** Fifteen Papers in Complex Analysis
145 **S. G. Dalalyan et al.,** Eight Papers Translated from the Russian
144 **S. D. Berman et al.,** Thirteen Papers Translated from the Russian
143 **V. A. Belonogov et al.,** Eight Papers Translated from the Russian
142 **M. B. Abalovich et al.,** Ten Papers Translated from the Russian
141 **H. Draškovičová et al.,** Ordered Sets and Lattices
140 **V. I. Bernik et al.,** Eleven Papers Translated from the Russian
139 **A. Ya. Aĭzenshtat et al.,** Nineteen Papers on Algebraic Semigroups
138 **I. V. Kovalishina and V. P. Potapov,** Seven Papers Translated from the Russian
137 **V. I. Arnol'd et al.,** Fourteen Papers Translated from the Russian
136 **L. A. Aksent'ev et al.,** Fourteen Papers Translated from the Russian
135 **S. N. Artemov et al.,** Six Papers in Logic
134 **A. Ya. Aĭzenshtat et al.,** Fourteen Papers Translated from the Russian

(Continued in the back of this publication)

Selected Papers
on Number Theory
and Algebraic Geometry

American Mathematical Society

TRANSLATIONS

Series 2 • Volume 172

Selected Papers on Number Theory and Algebraic Geometry

Katsumi Nomizu
Editor

American Mathematical Society
Providence, Rhode Island

Translation edited by KATSUMI NOMIZU

1991 *Mathematics Subject Classification.* Primary 11–XX, 14–XX

Library of Congress Cataloging-in-Publication Data
Selected papers on number theory and algebraic geometry / Katsumi Nomizu, editor.
  p. cm. — (American Mathematical Society translations, ISSN 0065-9290; ser. 2, v. 172)
 "A collection of five papers that originally appeared in the journal Sugaku from the Mathematical Society of Japan"—Pref.
 Includes bibliographical references.
 ISBN 0-8218-0445-6 (alk. paper)
 1. Number theory. 2. Geometry, Algebraic. I. Nomizu, Katsumi, 1924– . II. Sūgaku. III. Series.
QA3.A572 ser. 2, vol. 172
[QA241]
510 s—dc20
[512′.7]   95-39091
                                                                                                                                        CIP

Copying and reprinting. Material in this book may be reproduced by any means for educational and scientific purposes without fee or permission with the exception of reproduction by services that collect fees for delivery of documents and provided that the customary acknowledgment of the source is given. This consent does not extend to other kinds of copying for general distribution, for advertising or promotional purposes, or for resale. Requests for permission for commercial use of material should be addressed to the Assistant to the Publisher, American Mathematical Society, P. O. Box 6248, Providence, Rhode Island 02940-6248. Requests can also be made by e-mail to `reprint-permission@ams.org`.
 Excluded from these provisions is material in articles for which the author holds copyright. In such cases, requests for permission to use or reprint should be addressed directly to the author(s). (Copyright ownership is indicated in the notice in the lower right-hand corner of the first page of each article.)

© Copyright 1996 by the American Mathematical Society. All rights reserved.
The American Mathematical Society retains all rights
except those granted to the United States Government.
Printed in the United States of America.

∞ The paper used in this book is acid-free and falls within the guidelines
established to ensure permanence and durability.
♻ Printed on recycled paper.
10 9 8 7 6 5 4 3 2 1   01 99 98 97 96

Contents

Preface	ix
On Local Zeta Functions JUN-ICHI IGUSA	1
Mahler Functions and Transcendental Numbers KUMIKO NISHIOKA	21
Generalization of Class Field Theory KAZUYA KATO	31
Recent Topics on Open Algebraic Surfaces MASAYOSHI MIYANISHI	61
Recent Topics on Toric Varieties TADAO ODA	77

Preface

This is a collection of five papers that originally appeared in the journal Sugaku from the Mathematical Society of Japan. These articles in translation would normally appear in the AMS journal **Sugaku Expositions**. In order to expedite publication, however, we have made this selection and publish it as a volume of the Society's Translation Series II. We thank the Mathematical Society of Japan for their strong support of our translation program.

The articles in this volume range in 11 and 14 in the 1991 Mathematics Subject Classification. The following is no more than a brief indication of the content.

The main purpose of Igusa's paper is to study the Igusa zeta functions over p-adic fields. His first result in this area was a solution, obtained twenty years ago, of the Borewicz-Šafarevič conjecture.

Nishioka recalls a result of Lindemann-Weierstrass which says that if $\alpha_1, \alpha_2, \ldots, \alpha_m$ are algebraic numbers that are linearly independent over Q, then $e^{\alpha_1}, \ldots, e^{\alpha_m}$ are algebraically independent. The author gives a survey on various generalizations of this result including her own recent results on transcendance of function values.

Kato's paper studies various results obtained when one considers a generalization of class field theory for an algebraic field (namely, a finite extension of the field Q) to the case of a field generated by a finite number of elements over Q.

Miyanishi's paper first recalls the construction of the Ramanujam surface, a contractible homology plane that is not isomorphic to the affine plane $\mathbf{A}^2$. He then goes on to the classification of open algebraic surfaces and surveys other recent results, particularly, the work by the author, R. V. Gurjar, and D.-Q. Zhang.

Finally, Oda is concerned with the theory of toric varieties. The theory was developed in the 1970s as the theory that combines algebraic geometry with geometry of convex bodies and lattices. The paper documents the remarkable progress of the theory in conjunction with many important topics in mathematics of the last twenty years.

Katsumi Nomizu

On Local Zeta Functions

Jun-ichi Igusa

§0. Introduction

"Local" means a completion of an algebraic number field and "local zeta functions" are those that take complex values. In this article we shall mainly explain Igusa's zeta functions for p-adic fields. We have not restricted ourselves to the p-adic case because it is important to consider all completions to proceed in the investigation. Local zeta functions in the above sense are defined by using fundamental concepts in mathematics such as polynomials, local fields, and Haar measures. However, especially as some recent progress in the p-adic case shows, the results are not generalizations of known facts but are entirely new. The objective is to explore an area that might be called "mathematics of higher degree forms". We shall explain a general idea of the theory, postponing the details to the text.

The simplest Igusa's local zeta function is defined as follows. If K is a p-adic field, O_K is its ring of integers, $f(x)$ is a polynomial in n variables with coefficients in O_K, and dx is the Haar measure on K^n such that the total measure of O_K^n is 1, then Igusa's local zeta function is the function of a complex variable s defined as

$$Z(s) = \int_{O_K^n} |f(x)|_K^s \, dx, \quad \operatorname{Re}(s) > 0.$$

If we take reference [13] as the starting point of the theory, i.e., the reference in which the fundamental theorem stating that $Z(s)$ is a rational function of $t = q^{-s}$ with rational coefficients was proved, where q is the cardinality of the residue class field of O_K, and Borewicz-Šafarevič's conjecture was settled, then this volume marks exactly the 20th year since that paper appeared. Clearly the main problems on $Z(s)$ consist of examining its properties and, if possible, precisely describing its denominator and numerator as polynomials of t.

Concerning the properties of $Z(s)$, a new functional equation was experimentally discovered in the case where $f(x)$ is homogeneous and has a good reduction. This conjectural functional equation of $Z(s)$ was proved by Denef and Meuser [8] by using functional equations of Weil's zeta functions over F_q, in fact those for numerous varieties for just one $Z(s)$. This new type of functional equation was proved earlier for the zeta function of a connected irreducible K-split subgroup G of GL_n containing $\operatorname{GL}_1 \cdot 1_n$. As its corollary, the degree as a rational function of the variable of the

---

1991 *Mathematics Subject Classification.* Primary 11S31, 11S40, 14G10.

This article originally appeared in Japanese in Sūgaku **46** (1) (1994), 23–38.

©1996, American Mathematical Society

p-adic Hecke series of G was shown to be the negative of the order of the center of the commutator subgroup $D(G)$ of G. We might recall that this fact was a conjecture in the special case where $G = GSp_{2l}$ at the time when Satake developed his theory of p-adic spherical functions.

As for the denominator of $Z(s)$, the Bernstein polynomial $b_f(s)$ of $f(x)$, or simply the b-function, using the terminology in Sato's theory of prehomogeneous vector spaces, becomes important. For instance, there is a conjecture to the effect that the real part of a pole of $Z(s)$ of order k is necessarily a zero of $b_f(s)$ of multiplicity at least k. This conjecture is supported by a theorem on complex powers for $K = \mathbf{R}$ or $\mathbf{C}$ and a large number of examples in the p-adic case. It was proved in some special cases by Loeser [29], [30] and for the relative invariants of irreducible regular prehomogeneous vector spaces by Kimura, F. Sato, and Zhu [28] with no information about the orders of poles. In the general case it is open. As for the numerator of $Z(s)$, some examples in the prehomogeneous case are known such that certain cubic polynomials in t of the same type appear rather mysteriously in the $Z(s)$ for those $f(x)$ that have no similarity whatsoever. At any rate, this theory is still new and there are many conjectures and problems.

§1. Weil's conditions (A), (B)

Let G and X denote locally compact abelian groups, G^* the dual of G, and express elements of G, G^*, X and the Haar measures in G^*, X respectively by g, g^*, x and dg^*, dx. Further, put $\langle g, g^* \rangle = g^*(g)$ and denote by $\mathcal{S}(X)$ the Schwartz-Bruhat space of X. If now a continuous map $f : X \to G$ is given, for an arbitrary Φ in $\mathcal{S}(X)$ and g^* in G^* put

$$F_\Phi^*(g^*) = \int_X \Phi(x) \langle f(x), g^* \rangle \, dx.$$

As one can easily see, F_Φ^* is a bounded uniformly continuous function on G^*. In [43, pp. 6–7] Weil showed under the assumption that (A) F_Φ^* is an L^1-function on G^* for every Φ and the integral $\int |F_\Phi^*| \, dg^*$ is uniformly convergent on every compact subset of $\mathcal{S}(X)$ the existence of a unique family of tempered measures μ_g on X, each supported by $f^{-1}(g)$, such that the inverse Fourier transform F_Φ of F_Φ^* can be expressed as

$$F_\Phi(g) = \int_X \Phi(x) \, d\mu_g(x).$$

In general, if G contains a discrete subgroup Γ with compact quotient group G/Γ and if Γ_* denotes the annihilator of Γ in G^*, then Γ_* is a discrete subgroup of G^* with compact quotient group G^*/Γ_*. In such a case Weil showed in op. cit., pp. 7–8 the following fact. If (B) *the series*

$$\sum_{\gamma^* \in \Gamma_*} |F_\Phi^*(g^* + \gamma^*)|$$

is convergent for every Φ and g^* and the convergence is uniform on every compact subset of $\mathcal{S}(X) \times G^*$, then f satisfies Condition (A) and the formula

(P) $$\sum_{\gamma \in \Gamma} F_\Phi(\gamma) = \sum_{\gamma^* \in \Gamma_*} F_\Phi^*(\gamma^*)$$

holds with both sides absolutely convergent. This is a kind of Poisson formula, and it is an abstract form of the Siegel-Weil formula. In the latter part of the same paper Weil examined the above two conditions in the case where f is a quadratic map and converted (P) into what he called the "Siegel formula". After reading Weil's paper, the author became interested in the possibility of proving a Siegel-Weil formula for a higher-degree map. In such an attempt he faced the problem of clarifying the exact nature of (A), (B) and of converting them into verifiable conditions. When the author asked Weil about this problem in 1968 at Oberwolfach, Weil was fully aware of its importance and told the following story. Several years earlier he raised the question with Hörmander in the case where X and G are real vector spaces and f is a polynomial map, but Hörmander did not clarify the matter.

§2. Borewicz-Šafarevič's conjecture

For an arbitrary prime number p, denote by Z_p the p-adic completion of the ring of rational integers $\mathbf{Z}$; for any polynomial $f(x)$ in n variables $x_1, \ldots, x_n$ with coefficients in Z_p, denote by c_e for $e = 0, 1, 2, \ldots$ the number of solutions in Z_p^n, considered mod p^e, of $f(x) \equiv 0 \bmod p^e$. In order to examine the properties of c_e, introduce a power series

$$\varphi(t) = \sum_{e=0}^{\infty} c_e t^e$$

in a complex variable t. Clearly $\varphi(t)$ is convergent in the interior of the circle of radius p^{-n}; it was called the Poincaré series of $f(x)$ in [4, p. 63]. Furthermore, it was conjectured there that $\varphi(t)$ not only has a meromorphic continuation to the whole plane but also is a rational function of t. The author learned the existence of this conjecture from Dwork after proving a rationality theorem that included it. We have explained it here as an appropriate place.

§3. Complex power $\omega(f)$

We shall first explain the notation. We denote by $\mathbf{Q}, \mathbf{R}$, and $\mathbf{C}$ the fields of rational, real, and complex numbers. In general, if R is an associative ring with the unit element, then we denote by $R^\times$ the group of units of R. We choose a completion K of an algebraic number field; K is a locally compact field of characteristic 0 and it is $\mathbf{R}, \mathbf{C}$, or a p-adic field, i.e., a finite algebraic extension of the quotient field Q_p of Z_p for a certain prime number p. If a is an element of $K^\times$, we denote $d(ax)/dx$, the rate of change of a Haar measure dx on K under $x \to ax$, by $|a|_K$ and put $|0|_K = 0$. In this article we normalize dx as follows. The total measure of the compact subset of K defined by $|a|_K \leq 1$ is 2 (or 2π) for $K = \mathbf{R}$ (or $\mathbf{C}$) and is 1 if K is a p-adic field. We take the product measure on K^n as its Haar measure.

We take a arbitrarily from $K^\times$ and define its angular component $\mathrm{ac}(a)$. The subset $K_1^\times$ of $K^\times$ defined by $|a|_K = 1$ forms its maximal compact subgroup. In the case where $K = \mathbf{R}$ or $\mathbf{C}$, if we put $\mathrm{ac}(a) = a/|a|$, then $\mathrm{ac}(a)$ is an element of $K_1^\times$ and $a = |a| \cdot \mathrm{ac}(a)$. If K is a p-adic field, the subset O_K of K defined above as $|a|_K \leq 1$ forms a subring of K and $K_1^\times = O_K^\times$. Furthermore, $O_K - O_K^\times$ can be written as πO_K for some π in O_K, and if $q = |\pi|_K^{-1}$, then the factor ring $O_K/\pi O_K$ becomes the finite field F_q. If it becomes necessary to distinguish the above π from the other π, e.g., from 3.14..., then we shall denote it by π_K. Similarly we shall denote q by q_K

to avoid confusion. If now a is any element of $K^\times$, then there exist unique elements $\mathrm{ord}(a)$ and $\mathrm{ac}(a)$ of $\mathbf{Z}$ and $K_1^\times$ such that

$$a = \pi^{\mathrm{ord}(a)} \cdot \mathrm{ac}(a).$$

We denote by $\Omega(K^\times)$ the topological group of all continuous homomorphisms $\omega : K^\times \to \mathbf{C}^\times$. If for any complex number s we put

$$\omega_s(a) = |a|_K^s,$$

then we get an element ω_s of $\Omega(K^\times)$, and an arbitrary element ω of $\Omega(K^\times)$ can be written as

$$\omega(a) = \omega_s(a)\chi(\mathrm{ac}(a))$$

for some s, in which $\chi = \omega|K_1^\times$ is a character of $K_1^\times$. The character group of $K_1^\times$, i.e., its dual $(K_1^\times)^*$, is discrete, and the correspondence $\omega \to (s, \chi)$ for $K = \mathbf{R}$ or $\mathbf{C}$ and $\omega \to (q^{-s}, \chi)$ if K is a p-adic field give bicontinuous isomorphisms

$$\Omega(K^\times) \cong \mathbf{C} \times (K_1^\times)^*, \quad \mathbf{C}^\times \times (K_1^\times)^*.$$

If σ is any real number, we define an open subset $\Omega_\sigma(K^\times)$ of $\Omega(K^\times)$ as $\sigma(\omega) > \sigma$, where $\sigma(\omega)$ is the real part $\mathrm{Re}(s)$ of s. We call a function on $\Omega_\sigma(K^\times)$ holomorphic, etc. if it is holomorphic, etc. with respect to the local parameter s. In the p-adic case it is more appropriate to use $t = q^{-s}$ instead of s.

Now if a polynomial $f(x)$ in n variables with coefficients in K is given, we take elements Φ, ω of $\mathcal{S}(K^n), \Omega_0(K^\times)$ and consider the following integral:

$$Z_\Phi(\omega) = \int_{K^n} \Phi(x)\omega(f(x))\, dx$$

by the Haar measure dx on K^n. Then we get a holomorphic function $Z_\Phi(\omega)$ on $\Omega_0(K^\times)$. Furthermore, if we denote $Z_\Phi(\omega)$ by $\omega(f)(\Phi)$ and the topological dual of $\mathcal{S}(K^n)$ by $\mathcal{S}(K^n)'$, then $\omega(f)$ becomes an $\mathcal{S}(K^n)'$-valued holomorphic function, i.e., a holomorphic distribution, on $\Omega_0(K^\times)$ and is called the complex power of f. In particular, the complex power $\omega_s(f)$ is usually denoted by $|f|_K^s$ and has been investigated mainly by analysts for $K = \mathbf{R}$ or $\mathbf{C}$. For example, the problem of meromorphic continuation of complex powers proposed by I. M. Gel'fand at the Amsterdam Congress in 1954 was solved in 1968 by Bernstein and S. I. Gel'fand [2]. About this we also refer to Atiyah [1]. In the p-adic case we have the following theorem.

THEOREM 3.1. *If Φ is fixed arbitrarily in $\mathcal{S}(K^n)$, then $Z_\Phi(\omega)$ becomes a rational function of $t = \omega(\pi)$ for each χ. Furthermore, if the map $f : K^n \to K$ has no critical value in $K^\times$, i.e., if $f(C_f)$ for the critical set C_f of f is contained in $\{0\}$, then there exists a positive integer m such that $Z_\Phi(\omega) = 0$ identically for every χ satisfying $\chi|(1 + \pi^m O_K) \neq 1$.*

As we have already mentioned in the introduction, this theorem was proved by the author in 1973. Finally, $Z_\Phi(\omega)$, especially $Z(\omega) = Z_\Phi(\omega)$ for the characteristic function Φ of O_K^n, has often been called Igusa's local zeta function.

§4. Complex power and Condition (A)

Firstly, we shall explain the fact that the Borewicz-Šafarevič rationality conjecture is settled by Theorem 3.1. If the coefficients of $f(x)$ are contained in O_K, then we denote by c_e the number of solutions in O_K^n, considered mod π^e, of $f(x) \equiv 0 \bmod \pi^e$ and introduce the power series

$$P(t) = \sum_{e=0}^{\infty} c_e (q^{-n} t)^e.$$

Clearly $P(t)$ is convergent on the unit disc and if $t = \omega_s(\pi)$ for $\mathrm{Re}(s) > 0$, then we can easily see that

$$P(t) = (1 - tZ(\omega_s))/(1 - t).$$

Therefore, the rationality of $Z(\omega_s)$ in t implies the rationality of $P(t)$. If we take Q_p as K, then $\varphi(t)$ in §2 becomes $P(p^n t)$; hence $\varphi(t)$ is a rational function.

We shall now explain the relation between Condition (A) and the complex power. We take as K an arbitrary completion of an algebraic number field and assume that $f : K^n \to K$ for $f(x)$ in $K[x_1, \ldots, x_n]$ has no critical value in $K^\times$. This condition is satisfied if $f(x)$ is quasihomogeneous, hence if $f(x)$ is homogeneous. In order to state the next theorem uniformly in K, if K is a p-adic field, we restrict the imaginary part $\mathrm{Im}(s)$ of s as

$$-\pi/\log q \leq \mathrm{Im}(s) < \pi/\log q.$$

THEOREM 4.1. *If we take K^n and K as X and G, then Weil's condition (A) for $f : K^n \to K$ becomes equivalent to the following functions:*

$$(s+1)Z_\Phi(\omega), \; \chi = 1 \quad \text{and} \quad Z_\Phi(\omega), \; \chi \neq 1,$$

being holomorphic for $\mathrm{Re}(s) \geq -1$ for all Φ in $\mathscr{S}(K^n)$.

We refer to [13] for the proof. We might point out a formal analogy between the above condition on $Z_\Phi(\omega)$ and the well-known properties of Hecke's L-functions. If K is a p-adic field, the condition can also be stated as $Z_\Phi(\omega)$ being holomorphic for $0 < t \leq q$ except for $Z_\Phi(\omega_s)$ possibly having a pole of order 1 at $t = q$. At any rate, by the above theorem, the clarification of Condition (A) shifts to the investigation of the poles of complex powers.

§5. Poles of the complex power

The meromorphic continuation of $\omega(f)$ to the whole $\Omega(K^\times)$ was proved by using a desingularization theorem established by Hironaka [12]. The basic idea of the proof is as follows. Firstly, if we exclude the case where the element $f(x)$ of $K[x_1, \ldots, x_n]$ is contained in K, especially the trivial case where $f(x) = 0$, then there exist an everywhere n-dimensional smooth variety Y, a proper map $h : Y \to X = K^n$, and a finite set $\{E_i ; i \in I\}$ of closed subvarieties of Y with normal crossings, all K-analytic, such that the divisors of the composite function $f \circ h$ and the differential form $h^*(dx_1 \wedge \cdots \wedge dx_n)$ respectively become

$$\sum_{i \in I} N_i E_i, \quad \sum_{i \in I} (v_i - 1) E_i.$$

In the above statement h is surjective and every E_i is everywhere $(n-1)$-dimensional and smooth. Furthermore, N_i, v_i are positive integers and the set $\{(N_i, v_i); i \in I\}$

is called the numerical data of h. If, for the sake of uniformity, we define $\Gamma_K(s)$ as $\Gamma(s)$ or $\Gamma(2s)$ for $K = \mathbf{R}$ or $\mathbf{C}$ and $(1 - q^{-s})^{-1}$ for a p-adic field K, then we get a description of the poles of $\omega(f)$ by saying that

$$\prod_{i \in I} \Gamma_K(N_i s + v_i)^{-1} \cdot \omega(f)$$

is a holomorphic distribution on $\Omega(K^\times)$. In order to prove this fact, if we denote by μ the measure on Y defined by $h^*(dx_1 \wedge \cdots \wedge dx_n)$, then we have only to divide Y into small parts and examine the following integral on each part:

$$\int \Phi(h(y))\omega(f(h(y)))\, d\mu(y).$$

We see from the above description that the poles of $\omega(f)$ on the s-plane for $K = \mathbf{R}$ or $\mathbf{C}$ and the real parts of the poles of $\omega(f)$ also on the s-plane if K is a p-adic field are all negative rational numbers. Furthermore, the above-outlined proof shows that if $\{E_i; i \in J\}$ is a simplex of the nerve complex of $\{E_i; i \in I\}$, i.e., if J is a subset of I such that the intersection of E_i for all i in J is not empty, then every pole of $\omega(f)$ can be eliminated by the partial product of $\Gamma_K(N_i s + v_i)^{-1}$ for all i in a suitable J. Therefore, the order of the pole is at most n, in fact at most the dimension of the above nerve complex increased by 1.

We have seen that the poles of $\omega(f)$ can be described by the nerve complex of $\{E_i; i \in I\}$ and the numerical data defined on the set of its vertices. However, this has a drawback coming from the fact that the desingularization $h : Y \to X$ is not unique. Besides, the actual poles of $\omega(f)$ may form a very small part of the set of poles obtained as above. In the p-adic case this fact was first observed by Strauss for a minimal desingularization of a plane curve. We refer to Strauss [42] and [17, p. 220] for the details.

§6. b-function $b_f(s)$

In the case where $K = \mathbf{R}$ or $\mathbf{C}$ Bernstein gave a new proof to the meromorphic continuation of $\omega(f)$ by using a polynomial that is now called Bernstein's polynomial of $f(x)$. In the following we shall explain what it is. Let F_0 denote an arbitrary field of characteristic 0, $F = F_0(s)$ the field of rational functions in one variable s with coefficients in F_0 and A the associative F-algebra generated over F by commutative variables $x_1, \ldots, x_n$ and commutative symbols $\partial/\partial x_1, \ldots, \partial/\partial x_n$ with

$$x_i(\partial/\partial x_j) - (\partial/\partial x_j)x_i + \delta_{ij} = 0 \quad (1 \leq i, j \leq n)$$

as defining relations. Further, for any given element $f(x) \neq 0$ of $F_0[x_1, \ldots, x_n]$ introduce a symbol f^s and consider the module $M = F[x_1, \ldots, x_n, f(x)^{-1}]f^s$ over the ring of fractions $F[x_1, \ldots, x_n, f(x)^{-1}]$ of $F[x_1, \ldots, x_n]$. If now for an arbitrary element $\phi(x)f^s$ of M and for each i we define multiplications of x_i and $\partial/\partial x_i$ to $\phi(x)f^s$ as

$$x_i \cdot \phi(x)f^s = (x_i\phi(x))f^s,$$

$$\partial/\partial x_i \cdot \phi(x)f^s = (\partial\phi/\partial x_i + s\phi(x)f(x)^{-1}\partial f/\partial x_i)f^s,$$

in which $\partial\phi/\partial x_i, \partial f/\partial x_i$ are partial derivatives of ϕ, f with respect to x_i, then M becomes a left A-module and the following theorem holds.

THEOREM 6.1. *There exists an element Q of A satisfying $Q \cdot f(x) f^s = f^s$.*

For the proof we refer to Bernstein [3]. This theorem shows the existence of a polynomial P in s and the noncommutative letters $x_1, \ldots, x_n, \partial/\partial x_1, \ldots, \partial/\partial x_n$ with coefficients in F_0 such that

$$P \cdot f(x) f^s = b(s) f^s$$

for some $b(s) \neq 0$ in $F_0[s]$. The smallest-degree monic polynomial $b_f(s)$ among such $b(s)$ is uniquely determined by $f(x)$ and F_0, and this is the Bernstein polynomial of $f(x)$. In this article we shall use Sato's terminology and call $b_f(s)$ the b-function of $f(x)$. As one can easily see, $b_f(s)$ is invariant under any field extension of F_0, any multiplication by a nonzero constant, and any invertible affine linear transformation in $x_1, \ldots, x_n$. Furthermore, if we exclude the case where $f(x)$ is in F_0, then $b_f(s)$ has $s + 1$ as a factor, and if the hypersurface $f^{-1}(0)$ defined by $f(x)$ is smooth, then $b_f(s) = s + 1$. Therefore, in the general case, other factors of $b_f(s)$ are caused by the singularity of $f^{-1}(0)$. At any rate if we write

$$b_f(s) = \prod_\lambda (s + \lambda),$$

then by Kashiwara [26] all λ's are positive rational numbers.

§7. b-function and the complex power

The mechanism of Theorem 6.1 for $F_0 = K = \mathbf{R}$ or $\mathbf{C}$ implying the meromorphic continuation of $\omega(f)$ is very simple, and in the case where $F_0 = \mathbf{R}$ it is as follows. The set of real points $f^{-1}(0)_\mathbf{R}$ of the hypersurface $f^{-1}(0)$ divides $\mathbf{R}^n$ into a finite number of connected components. If Θ is a union of some such components on which $f > 0$ and if for an arbitrary Φ in $\mathcal{S}(\mathbf{R}^n)$ and $\text{Re}(s) > 0$ we put

$$f_\Theta^s(\Phi) = \int_\Theta \Phi(x) f(x)^s \, dx,$$

then f_Θ^s becomes a holomorphic distribution on the right half-plane. If we regard P as a differential operator and denote its adjoint by P^*, then $P^*\Phi$ is also an element of $\mathcal{S}(\mathbf{R}^n)$ and we get

$$b_f(s) f_\Theta^s(\Phi) = f_\Theta^{s+1}(P^*\Phi).$$

In order to prove this we remark that the identity

$$P f(x)^{s+1} = b_f(s) f(x)^s$$

holds at every point x of Θ with P as a differential operator and we apply integration-by-parts and Stokes' theorem repeatedly for $\text{Re}(s)$ sufficiently large. The above identity shows that $b_f(s) f_\Theta^s$ is a holomorphic distribution for $\text{Re}(s) > -1$ and, therefore, if we put

$$b_f(s) = \prod_\lambda (s + \lambda), \qquad \gamma_f(s) = \prod_\lambda \Gamma(s + \lambda),$$

then $\gamma_f(s)^{-1} \cdot f_\Theta^s$ becomes a holomorphic distribution on the whole s-plane. In particular, if we take the largest Θ as above, denote the corresponding f_Θ^s by f_+^s and put $f_-^s = (-f)_+^s$, then we see that $\gamma_f(s)^{-1} \cdot \omega(f)$ is holomorphic on $\Omega(\mathbf{R}^\times)$, this by using $\omega(f) = f_+^s \pm f_-^s$ and $b_{-f}(s) = b_f(s)$. The case where $K = \mathbf{C}$ is similar.

If we define an integer m as $\omega(a/|a|) = (a/|a|)^m$, put $s_1 = s + m/2$, $s_2 = s - m/2$ and examine

$$\omega(f)(\Phi) = f^{s_1}\overline{f}^{s_2}(\Phi), \quad \Phi \in \mathcal{S}(\mathbf{C}^n)$$

in the same way, then we see that $\gamma_f(s + |m|/2)^{-1} \cdot \omega(f)$ is a holomorphic distribution on $\Omega(\mathbf{C}^\times)$. Therefore, the investigation of the poles of $\omega(f)$ shifts to the problem of determining $b_f(s)$. At any rate if we take ω_s as ω, then the above results imply the following theorem.

THEOREM 7.1. *If $K = \mathbf{R}$ or $\mathbf{C}$ and the λ's are as above, then*

$$\prod_\lambda \Gamma(s + \lambda)^{-1} \cdot |f|_K^s$$

is a holomorphic distribution on the whole s-plane.

Now in the case where K is a p-adic field the above proof breaks down completely. Nevertheless, in view of the results in §5 and by a large number of examples, an analogous statement becomes the following.

CONJECTURE 7.2. *In the case where K is a p-adic field, the real parts of the poles of $|f|_K^s$ are roots of $b_f(s)$ and the order of each pole does not exceed the multiplicity of the root of $b_f(s)$.*

Since every element of $\mathcal{S}(K^n)$ is a $\mathbf{C}$-linear combination of the transforms of the characteristic function of O_K^n under invertible affine linear transformations, we can replace $|f|_K^s$ by

$$Z(s) = \int_{O_K^n} |f(x)|_K^s \, dx.$$

We have explained the so-reformulated conjecture already in the introduction.

§8. Serre-Oesterlé's conjecture

We go back to the Borewicz-Šafarevič conjecture for an element $f(x)$ of $Z_p[x_1, \ldots, x_n]$. They considered the rationality of the power series in t with the number c_e of solutions of $f(x) \equiv 0 \bmod p^e$ in Z_p^n, considered mod p^e, as the coefficient of t^e for $e = 0, 1, 2, \ldots$. In [40, p. 146] Serre asked a similar question by using the number of solutions of $f(x) = 0$ in Z_p^n, considered mod p^e, instead of the c_e above. Furthermore, in [33, p. 326] Oesterlé considered a closed Q_p-analytic subvariety V of Z_p^n, denoted by V_e the image of V under the canonical map $Z_p^n \to (Z_p/p^e Z_p)^n$, and proposed the question of rationality of the power series

$$\varphi^\sharp(t) = \sum_{e=0}^\infty \text{card}(V_e) t^e,$$

where "card" stands for the cardinality, as Serre's problem. This problem was settled in the algebraic case, i.e., in the case where V is algebraic, by Denef [5] and in the above analytic case by Denef and van den Dries [7]. In the following we shall explain Denef's solution.

Take a finite algebraic extension K of Q_p, denote elements of $K[x_1, \ldots, x_n]$ by $g(x), h(x), \ldots$ and assume that $m = 2, 3, \ldots$. Define subsets $A_g, B_{g,h}, C_{g,m}$ of K^n respectively by the following conditions: $g(x) = 0$, $|g(x)|_K \le |h(x)|_K$, $g(x) = y^m$ for some element y of K. Consider the family of $A_g, B_{g,h}, C_{g,m}$ for all $g(x)$, all $g(x), h(x)$, all $g(x)$, m and apply the operations of taking unions, intersections, and complements finitely many times to the members of the family. Call subsets of K^n obtained in that way semialgebraic. Then by Macintyre [32] the image of any semialgebraic subset of $K^n \times K^{n'}$ under the projection $K^n \times K^{n'} \to K^n$ is a semialgebraic subset of K^n. By using this fact Denef succeeded in proving the following theorem.

THEOREM 8.1. *If S is a semialgebraic subset of K^n that is bounded, i.e., with compact closure, then for any $f(x)$ in $K[x_1, \ldots, x_n]$ and for $\mathrm{Re}(s) > 0$ the integral $\int_S |f(x)|_K^s\, dx$ is a rational function of $t = q^{-s}$.*

This theorem clearly implies the rationality of $Z_\Phi(\omega)$. Also, Denef derived the algebraic case of Serre-Oesterlé's conjecture from his theorem as follows. In general, for any semialgebraic subset V of K^n contained in O_K^n consider the set of all elements (x, x', y) of $O_K^n \times V \times O_K$ satisfying $x \equiv x' \bmod y$ and denote by W its image under $(x, x', y) \to (x, y)$. Then by Macintyre's theorem W is a semialgebraic subset of K^{n+1} and it is clearly bounded. Furthermore, if we put

$$P^\sharp(t) = \sum_{e=0}^\infty \mathrm{card}(V_e)(q^{-n}t)^e$$

and $t = q^{-s}$ for $\mathrm{Re}(s) > 0$, then we can easily see that

$$\int_W |y|_K^s\, dx\, dy = (1 - q^{-1})P^\sharp(q^{-1}t).$$

The left-hand side is a rational function of t by Theorem 8.1; hence $P^\sharp(t)$ is a rational function and, in particular, $\varphi^\sharp(t) = P^\sharp(p^n t)$ for $K = Q_p$ is a rational function. We might mention that in [5] Denef first proved his theorem by using Hironaka's desingularization theorem and then gave another proof without using that theorem.

§9. Prehomogeneous vector spaces

We have been following, up to this point, the history of meromorphic continuations of local zeta functions and their rationality in the p-adic case. Meanwhile, the author examined many examples of p-adic zeta functions to explore the possibility of finding new theorems. More precisely, the author systematically examined the $Z(s)$ for the relative invariant $f(x)$ of an irreducible regular prehomogeneous vector space, abbreviated as IRPVS. We shall first recall the definition of IRPVS referring to Sato [36] and Sato and Kimura [38] for the details.

Take a field F of characteristic 0, denote the n-dimensional affine space over F by Aff^n, the group of invertible linear transformations in Aff^n by GL_n and put $X = \mathrm{Aff}^n$ so that $X(F) = X_F = F^n$. If a connected irreducible algebraic subgroup G of GL_n acts transitively on the complement Y of a hypersurface in X, then the pair (G, X) is called an IRPVS. In this case if we denote by $D(G)$ the commutator subgroup of G and by 1_n the unit element of GL_n, then $D(G)$ is semisimple and G is a product of $D(G)$ and $\mathrm{GL}_1 \cdot 1_n$ such that their intersection is the center of $D(G)$. Furthermore, the hypersurface $X - Y$ can be written as $f^{-1}(0)$ with an irreducible relative invariant $f(x)$ of G which is unique up to a constant factor. If G is defined over F, we may assume that the coefficients of $f(x)$ are in F.

In the following we shall assume that G's are all defined over F and identify (G, X) and (G', X) if $G' = hGh^{-1}$ for some h in $\mathrm{GL}_n(F)$. We represent (G, X) by a point in the space with the following convention. We put the point representing

(G, X) over the point representing (G', X') if $\dim(X) > \dim(X')$. We then join two points representing $(G, X), (G', X')$ if they are mutually the "castling transforms" as defined in [38, p. 39]. In that way we get infinitely many connected components and each looks like a "tree". An IRPVS is called reduced if it is represented by the root of a tree, i.e., by a lowest point. In op. cit. Sato and Kimura classified for $F = \mathbf{C}$ the so-defined set of trees into 24 individual trees, 3 "rows" of trees, and 2 "forests", 29 types altogether; further, they gave a list of reduced IRPVS with their relative invariants. If we consider only those G's that are split over F, then Sato-Kimura's classification holds for any F.

Now in order to determine the b-function of a relative invariant $f(x)$, since the manner in which $b_f(s)$ changes under the castling transformation is known, cf., [27, p. 78], we may assume that (G, X) is reduced. Furthermore, as we shall later explain, if $'G = G$, then

$$f(\partial/\partial x) \cdot f(x) f^s = b_0 b_f(s) f^s$$

for some constant $b_0 \neq 0$. The problem of determining the b-function was settled by Kashiwara, Kimura, Muro, Oshima, Ozeki, Sato, and Yano. Pertaining to this we refer to the fundamental paper [37] by Sato and others as well as [27] and [44]. In Kimura [27] a list of b-functions of all reduced IRPVS, except for the most difficult case determined later by Yano and Ozeki [44], is given with references.

§10. Examples of $Z(s)$

We take a p-adic field K for $p \neq 2$, i.e., for q odd. If $f(x)$ is an arbitrary quadratic form with coefficients in K, then a fairly concise closed form for the corresponding $Z(s)$ is given in [24]. If the degree of $f(x)$ is higher, except for simple cases such as determinants, the computation of $Z(s)$ becomes very difficult. The first case in which the author felt such a difficulty was the case of a Freudenthal quartic, which is as follows.

We take a composition algebra C defined over K and denote by A the Jordan algebra of Hermitian matrices of degree three with coefficients in C. We denote the generic trace and norm of an element a of A by $T(a)$ and $N(a)$, and define a quadratic map $a \to a^\sharp$ as $aa^\sharp = N(a)1_3$. If we put $X = \text{Aff}^2 \times A^2$, then the Freudenthal quartic is defined for $x = (\alpha, \beta; a, b)$ in X as

$$f(x) = (\alpha\beta - T(ab))^2 + 4(\alpha N(b) + \beta N(a) - T(a^\sharp b^\sharp)).$$

We shall write down the corresponding $Z(s)$ after introducing the following irreducible polynomial $C_{m,n}(u, v)$ in two variables u, v, in which m, n are integers satisfying $m \geq n \geq 1$:

$$C_{m,n}(u, v) = (1 + u^m) - (1 + u^{m-n} + u^{2m-2n} - u^{2m-n})u^{n+1}v$$
$$+ (1 - u^n - u^m - u^{2m-n})u^{m+1}v^2 + (1 + u^m)u^{2m+2}v^3.$$

Also, for natural numbers m, n and a complex variable s we put

$$[m, n] = 1 - q^{-(m+ns)}, \qquad [m] = [m, 0].$$

If now C splits over K and $2k = \dim(C) \neq 1$, hence $k = 1, 2, 4$, then

$$Z(s) = [1][3k+2]C_{3k+2,2k+2}(q^{-1}, q^{-s})/[1,1][2k+3,2][4k+3,2][6k+4,2].$$

A proof for $k = 4$ is given in [15]. Other cases can be proved similarly; cf. [16]. If we take the product of $s + m/n$ for each factor $[m, n]$ of the denominator of $Z(s)$, then we get

$$b_f(s) = (s + 1)(s + k + \tfrac{3}{2})(s + 2k + \tfrac{3}{2})(s + 3k + 2),$$

which is indeed the b-function of $f(x)$ by [27] including the case where $k = 1/2$.

We shall explain one more example. This time we denote by X the space $M_{m,n}$ of $m \times n$ matrices, where $m \geq 2n$, and take a symmetric matrix h from $M_m(O_K)$ such that $(-1)^{m(m-1)/2} \det(h)$ is contained in $(O_K^\times)^2$. If for x in X we put

$$f(x) = \det({}^t x h x),$$

the corresponding $Z(s)$ is computed in [21]. The fact is that if m, n are both even and only in that case, for some unknown reason, the above-introduced cubic polynomial in $t = q^{-s}$ appears in the numerator:

$$Z(s) = [m/2]C_{m/2,n}(q^{-1}, q^{-s})/[n][1,1][n+1,2]$$

$$\cdot \prod_{i=1}^{n}[i]/[m-i+1,2] \cdot \prod_{j=1}^{n/2-1}[m-2j][m-2j+1,2]/[2j][2j+1,2].$$

In this example the same procedure as above will give only a part of the b-function of $f(x)$, which is

$$b_f(s) = \prod_{i=1}^{n}(s + (i+1)/2)(s + (m-i+1)/2)$$

by [37] without the assumption that m, n are both even.

We shall not give any more concrete examples, but mention that among the 29 types of IRPVS in Sato-Kimura's classification, if we exclude 5 trees, then $Z(s)$ has been computed for the relative invariant of each of the remaining 24 types. Since the manner in which $Z(s)$ changes under the castling transformation is known, cf. [19, p. 226], we have only to consider reduced cases. In [20] a list of $Z(s)$ for 20 types is given and the $Z(s)$ for the additional 4 types are determined in [22], [23].

§11. New functional equation (for $(Z(s))$

We take a close look at the two examples in §10. In the first example, $f(x)$ is a degree 4 form and the degree of $Z(s)$ as a rational function of $t = q^{-s}$ is -4 and, in the second example, $f(x)$ is a degree $2n$ form and the degree of $Z(s)$ in t is $-2n$. Since we observed this fact in many examples known at the time when [16] was written, on p. 1028 of that paper we proposed the following theorem as a problem.

THEOREM 11.1. *If a homogeneous polynomial $f(x)$ with coefficients in O_K has a good reduction mod π, then*

$$\deg_t(Z(s)) + \deg(f) = 0.$$

Furthermore, in that paper, we gave the following formulation without using "good reduction". If the coefficients of $f(x)$ are contained in an algebraic number field k, then the above relation holds for almost all p-adic completions K of k. At any rate the above theorem was proved by Meuser in a special case and by Denef in the general case with the understanding that "good reduction" means the existence of a desingularization over O_K. We refer to [6] for the details.

We return to the examples in §10. In each case $Z(s)$ is clearly a rational function of q^{-1} and q^{-s} such that the coefficients do not involve any other letter. Therefore, the operation of replacing q^{-1} and q^{-s} respectively by q and q^s, i.e., of replacing q by q^{-1}, has a well-defined meaning. Under that operation the first $Z(s)$ is multiplied by q^{-4s} and the second $Z(s)$ by q^{-2ns}. We examined this remarkable fact in all examples known in 1987 when we wrote a preliminary form of the first part of [21] and proposed the following theorem as a conjecture.

THEOREM 11.2. *Suppose that $f(x)$ satisfies the same condition as in the previous theorem and suppose further that there exists a rational function $Z(u,v)$ of two variables u,v such that if $Z_L(s)$ denotes the $Z(s)$ for the same $f(x)$ relative to a finite algebraic extension L of K, then*
$$Z_L(s) = Z(q_L^{-1}, q_L^{-s})$$
for all L. Then the unique $Z(u,v)$ satisfies the functional equation
$$Z(u^{-1}, v^{-1}) = v^{\deg(f)} Z(u,v).$$

The above conjecture was investigated by Meuser and it became a theorem with Denef's cooperation. Since their proof is instructive, we shall explain some key points, referring to [8] for the details. In §5 we used a K-analytic desingularization of $f^{-1}(0)_K$. If $f(x)$ is homogeneous and has a good reduction mod π, then there exists an O_K-desingularization of the hypersurface in the $(n-1)$-dimensional projective space defined by $f^{-1}(0)$ with similar properties. Namely, Y is an $(n-1)$-dimensional closed smooth projective variety and E_i for i in I are closed subvarieties of Y with normal crossings, etc. If d is the degree of $f(x)$ and if E_J for a subset J of I denotes the intersection of E_i for all i in J, then the following important formula:
$$Z(s) = q^{-n+1}[1]/[n,d] \cdot \sum_J \mathrm{card}(E_J(F_q)) \prod_{i \in J}(-[v_i - 1, N_i]/[v_i, N_i]),$$
called Denef's formula, holds. A similar formula holds more generally for $Z(\omega)$ and it has been used as starting points of various investigations. For example, Theorem 11.1 follows immediately from
$$\lim_{\mathrm{Re}(s) \to -\infty} q^{-ds} Z(s) = (1-q) \cdot \sum_J \mathrm{card}(E_J(F_q))(-q)^{\mathrm{card}(J)}$$
$$\equiv \mathrm{card}(Y(F_q)) \equiv 1 \bmod q.$$

As for the proof of Theorem 11.2, if $\overline{E}_J$ denotes $E_J \bmod \pi$, then $\overline{E}_J$ is a smooth projective F_q-variety; hence, the Poincaré duality by Grothendieck, i.e., the functional equation of its Weil zeta function, holds. The examination of $\mathrm{card}(E_J(F_q))$ in Denef's formula by using this duality with all q replaced by q^e for $e = 1, 2, 3, \ldots$ implies the new functional equation. Actually, if instead of $Z(u,v)$ we allow a rational function $Z(u_1, u_2, \ldots, v)$ of several variables, then it is known that a similar functional equation holds under the condition of Theorem 11.1 only.

§12. New functional equation (algebraic-group case)

Firstly, we shall recall Serre's canonical measure μ_c on an everywhere n-dimensional K-analytic submanifold X of K^N, where K is a p-adic field. It is the unique measure such that if $T_a(X)$ denotes the tangent space of X at a, then the total measure of the intersection of $T_a(X)$ and O_K^N determined by μ_c is 1 for every a in X. We refer to [40, p. 146] for the details. By definition μ_c is invariant under the stabilizer of X in $\mathrm{GL}_N(O_K)$ and also it is a generalization of the normalized Haar measure on K^n because it clearly reduces to that for $X = K^n$, $N = n$. Now in Part I of [21] we formulated a conjectural functional equation for a generalization of $Z(s)$ defined by μ_c and, in order to explore the domain of validity of such a functional equation, we examined in Part II the p-adic zeta function of an algebraic group. In the following we shall outline some of its aspects.

We take a connected irreducible K-split algebraic subgroup G of GL_n containing $\text{GL}_1 \cdot 1_n$. Then the coordinate ring of the closure of G in M_n is homogeneous and $\text{Hom}(G, \text{GL}_1)$, the group of morphisms from G to GL_1, is generated by its unique homogeneous element f of degree m, where m is the order of the center of the commutator subgroup $D(G)$ of G. We choose a maximal K-split torus T of G, denote by W and R respectively the Weyl group and the root system of G, and we choose a basis $S = (\alpha_1, \ldots, \alpha_l)$ for R. We denote by R^+ and C respectively the set of positive roots and the positive Weyl chamber relative to S, and we put

$$\Delta_S(t) = \prod_{\alpha \in R^+} |\alpha(t)|_K^{-1}$$

for every t in T_K. Also we denote by G^0 the intersection of G_K and $M_n(O_K)$, and by Ξ^0 the set of all elements ξ of $\Xi = \text{Hom}(\text{GL}_1, T)$ such that $\xi(\pi)$ is contained in G^0. Finally, we denote by μ the Haar measure on G_K that restricts to the canonical measure on $G(O_K)$ and put

$$\mu_0(g) = |f(g)|_K^{\delta} \mu(g), \qquad \delta = \dim(G)/m.$$

We recall that $G(O_K)$ is the subset of G^0 defined by $\det(g) \not\equiv 0 \bmod \pi$, and it is a compact open subgroup of G_K. The following theorem was proved under the assumption that G and its K-splitting data have good reductions mod π in the usual sense.

THEOREM 12.1. *Let Φ denote a complex-valued $K^\times G(O_K)$-bi-invariant function on G_K such that*

$$\Phi(\xi(\pi)) = \Delta_S(\xi(\pi))^{-\kappa} \cdot \sum_{w \in W} \gamma_{S,w}(s) w(s)(\xi(\pi))$$

for every ξ in the intersection of Ξ^0 and the closure of C, in which κ is in $\mathbf{C}$, s is a continuous homomorphism from $T_K/K^\times T(O_K)$ to $\mathbf{C}^\times$ and, if we write

$$s(t) = \prod_{i=1}^{l} |\alpha_i(t)|_K^{s_i},$$

then $\gamma_{S,w}(s)$ is of the form $\gamma_{S,w}(q^{-1}, q^{-s_1}, \ldots, q^{-s_l})$ for some rational function $\gamma_{S,w}(u_0, u_1, \ldots, u_l)$ of $l + 1$ variables $u_0, u_1, \ldots, u_l$. Then, for a complex variable s_0, the following integral

$$Z(s_0, \Phi) = \int_{G^0} \Phi(g) |f(g)|_K^{s_0} d\mu_0(g)$$

is absolutely convergent for all large $\text{Re}(s_0)$ and becomes a rational function of $q^{-1/m}$,

$q^{-\kappa}, q^{-s_0}, \ldots, q^{-s_l}$. Furthermore, if $\gamma_{S,w}(s)$ is invariant under the substitution $q \to q^{-1}$, then the zeta function $Z(s_0, \Phi)$ satisfies the functional equation

$$Z(s_0, \Phi)|_{q \to q^{-1}} = q^{-ms_0} \cdot Z(s_0, \Phi);$$

if further $\Phi(1_n) \neq 0$, then

$$\deg_v(Z(s_0, \Phi)) + m = 0, \quad v = q^{-s_0}.$$

The above theorem is stated by using a Cartan decomposition of G_K, but in its proof an Iwahori decomposition [25] was used. We refer to [21, p. 707] for an explicit form of $Z(s_0, \Phi)$ and to p. 709 for the argument to derive its degree from the functional equation. If now we denote by ϖ_s the spherical function associated with s as in Satake [35] and put $\Phi(g) = \varpi_s(g^{-1})$, then we see by a result of MacDonald [31] that Φ satisfies all conditions in the theorem. Therefore we get the following result.

COROLLARY 12.2. *Denote by G_e^0 the set of all elements g of G^0 satisfying $\mathrm{ord}(f(g)) = e$ and by τ_e the characteristic function of G_e^0 for $e = 0, 1, 2, \ldots$. Then the power series*

$$H(v) = \sum_{e=0}^{\infty} \tau_e v^e$$

with coefficients in the Hecke ring of $(G_K, G(O_K))$ is a rational function of v and its degree is equal to $-m$.

The above statement on the degree of $H(v)$ was stated as a conjecture in Satake, op. cit., p. 63 in the special case where $G = GSp_{2l}$. As we have mentioned in the introduction, the fact that it can be derived in the above general form rather naturally from the new functional equation seems to be remarkable.

§13. Γ-matrix $\Gamma(\omega)$

We go back to §9, denote by K any completion of an algebraic number field, and explain the classical functional equation of the relative invariant $f(x)$ of a prehomogeneous vector space defined over K. If we keep on taking Aff^n as X, then the dual space X^* of X cannot be clearly distinguished from X. Therefore, we take as X an affine K-space, i.e., an affine space defined over K, and put $[x, x^*] = x^*(x)$ for x, x^* in X, X^*. We choose a continuous homomorphism $\psi \neq 0$ from K to $C_1^\times$, put $\langle x, x^* \rangle = \psi([x, x^*])$ for x, x^* in X_K, X_K^* and denote by dx, dx^* dual measures on them. We recall that dx, dx^* are Haar measures on X_K, X_K^* such that the Fourier transformation $\mathscr{F} : \mathcal{S}(X_K) \to \mathcal{S}(X_K^*)$ defined by

$$\mathscr{F}(\Phi)(x^*) = \int_{X_K} \Phi(x) \langle x, x^* \rangle \, dx$$

is L^2-norm preserving. They are not unique, but their product measure is unique. We also recall that the group $\mathrm{GL}(X)$ of invertible linear transformations in X acts on X^* as $g \cdot x^* = {}^t g^{-1} x^*$, in which ${}^t g$ is defined as $[gx, x^*] = [x, {}^t g x^*]$.

Now suppose that $f(x)$ is an irreducible polynomial on X and that G is a connected reductive algebraic K-subgroup of $\mathrm{GL}(X)$ acting transitively on $Y = X - f^{-1}(0)$. Then there exists an irreducible polynomial $f^*(x^*)$ on X^* such that G also acts transitively on $Y^* = X^* - (f^*)^{-1}(0)$, and we may assume, after multiplying nonzero constants, that $f(x), f^*(x^*)$ are K-polynomials. Furthermore, there exists a unique element v of $\mathrm{Hom}(G, \mathrm{GL}_1)$, defined over K, such that

$$f(gx) = v(g)f(x), \qquad f^*(g \cdot x^*) = v(g)^{-1}f^*(x^*)$$

for every g in G. We know that $f(x), f^*(x^*)$ are homogeneous of the same degree, say d, and that

$$f^*(\partial/\partial x) \cdot f(x)f^s = b_0 b_f(s)f^s, \quad \deg(b_f(s)) = d$$

for some b_0 in $K^\times$, and that the correspondence $x \to f(x)^{-1}\partial f/\partial x$ gives a G-equivariant K-isomorphism $\varphi: Y \cong Y^*$. Since G is transitive on Y, the set $G_K \setminus Y_K$ of G_K-orbits in Y_K is finite by Serre [39, III-33] and if $l = \mathrm{card}(G_K \setminus Y_K)$, then

$$G_K \setminus Y_K = \coprod_{i=1}^{l} Y_i, \qquad G_K \setminus Y_K^* = \coprod_{i=1}^{l} Y_i^*,$$

in which $Y_i^* = \varphi(Y_i)$ for $1 \le i \le l$. Furthermore, if we put

$$d\mu(x) = \omega_{-\kappa}(f(x))\,dx, \qquad d\mu^*(x^*) = \omega_{-\kappa}(f^*(x^*))\,dx^*,$$

where $\kappa = \dim(X)/d$, then μ, μ^* give G_K-invariant measures on the homogeneous spaces Y_i, Y_i^*. Finally, we take ω from $\Omega_\kappa(K^\times)$ and for arbitrary elements Φ, Ψ of $\mathcal{S}(X_K), \mathcal{S}(X_K^*)$ we put

$$Z_i(\omega)(\Phi) = \int_{Y_i} \Phi(x)\omega(f(x))\,d\mu(x), \qquad Z_i^*(\omega)(\Psi) = \int_{Y_i^*} \Psi(x^*)\omega(f^*(x^*))\,d\mu^*(x^*).$$

Then we have the following theorem.

THEOREM 13.1. *The $2l$ holomorphic distributions $Z_i(\omega), Z_i^*(\omega)$ on $\Omega_\kappa(K^\times)$ all have meromorphic continuations to $\Omega(K^\times)$ and for every Φ in $\mathcal{S}(X_K)$ they satisfy a system of functional equations of the form*

$$Z_i^*(\omega)(\mathcal{F}(\Phi)) = \sum_{j=1}^{l} \gamma_{ij}(\omega) Z_j(\omega_\kappa \omega^{-1})(\Phi), \quad 1 \le i \le l,$$

in which the l^2 coefficients $\gamma_{ij}(\omega)$ are meromorphic functions on $\Omega(K^\times)$. If K is a p-adic field, then $Z_i(\omega)(\Phi)$ and $\gamma_{ij}(\omega)$ are all rational functions of $t = \omega(\pi)$.

This theorem was proved for $K = \mathbf{R}$ or $\mathbf{C}$ by Sato [36]; cf. also Shintani [41]. If K is a p-adic field, it was proved by the author [16] under the assumption that the number $\mathrm{Ker}(v)$-orbits in $f^{-1}(0)$ is finite. In the proof Denef's theorem, Theorem 8.1, was used. Recently, Gyoja [11] succeeded in removing the above finiteness assumption. The square matrix $\Gamma(\omega)$ with $\gamma_{ij}(\omega)$ as its (i, j)-entry is called a Γ-matrix. In the special case in which $X = X^* = \mathrm{Aff}^1, [x, x^*] = xx^*, dx = dx^*, G = \mathrm{GL}_1, f(x) = x$, $f^*(x^*) = x^*$, hence $l = 1$, the Γ-matrix is well known in the Tate theory; we shall denote it by $\gamma(\omega, \psi)$.

The above $\Gamma(\omega)$ contains too much freedom. Therefore, in [18] the author introduced a normalized Γ-matirx $a_K(G, \omega)$ as follows. Firstly, choose a K-isomorphism $h: X^* \cong X$ satisfying $G = h^t G h^{-1}$. This is always possible for $K = \mathbf{R}, \mathbf{C}$. In the case in which K is a p-adic field the existence of such an h is known, e.g., if G

is irreducible and splits over K. At any rate, once we choose h, then we can take $f(hx^*)$ as $f^*(x^*)$. Furthermore, we normalize dx so that

$$\Phi(x) \to \Phi^*(y) = \int_{X_K} \Phi(x)\langle x, h^{-1}y\rangle\, dx$$

becomes L^2-norm preserving. Then

$$a_K(G,\omega) = (\omega_{\kappa/2}\omega^{-1})(b_0) \prod_\lambda \gamma(\omega\omega_{\lambda-\kappa}, \psi)^{-1} \cdot \Gamma(\omega),$$

where $b_f(s) = \prod(s+\lambda)$ as always, depends only on (G,X) and the ordering of $Y_1, \ldots, Y_l$.

§14. Structure of $\Gamma(\omega)$

It is important for some applications to make the functional equation in Theorem 13.1 explicit, i.e., to determine $\Gamma(\omega)$. Since Tate's local factors are known, we have only to determine $a_K(G,\omega)$. Firstly, we have the following theorem.

THEOREM 14.1. $a_\mathbf{C}(G,\omega) = 1$.

This theorem was proved with ± 1 on the right-hand side by Sato [36] and in the above form by the author [18]. The next theorem is rather special, but it holds for every K.

THEOREM 14.2. *If all roots of $b_f(s)$ for the relative invariant $f(x)$ of an IRPVS (G,X) are integers and G splits over K, then $l = 1$ and $a_K(G,\omega) = 1$.*

Furthermore, even if G does not split over K, if the twisting is inner and $l = 1$, then there is a conjecture stating that X can be written as a direct sum of central division K-algebras D_i such that

$$a_K(G,\omega) = \prod_i (-1)^{d_i - 1}, \quad d_i^2 = \dim(D_i).$$

For the proof of Theorem 14.2 and this conjecture, we refer to [18] and Pan [34].

If $K = \mathbf{R}$ but is otherwise general, then $\Gamma(\omega)$ is known fairly explicitly. In fact, up to Fourier polynomials in s of limited degree, the entries of $\Gamma(\omega)$ are determined in [36], [41]. In the case in which K is a p-adic field we have the following theorem.

THEOREM 14.3. *If one of $Z_i(\omega), \gamma_{ij}(\omega)$ has ϖ in $\Omega(K^\times)$ as a pole, then there exists a point ξ in $f^{-1}(0)_K$ such that for the fixer or the stabilizer H of ξ in G_K*

$$\varpi \circ \nu | H = \Delta_H \ (= \text{the module of } H)$$

holds.

We refer to [16] and Gyoja [11] for the proof. We might mention that in the proof of the result by Kimura, F. Sato, and Zhu [28] concerning the conjecture 7.2, the above theorem and a theorem of Sato are used.

In the case in which K is a p-adic field we can also normalize $\Gamma(\omega)$ as follows. Firstly, by using dual bases for X_K, X_K^* we replace X, X^* by Aff^n and take the normalized Haar measure on K^n as dx, dx^*. We then normalize ψ by the condition that $\mathscr{F}$ is L^2-norm preserving, i.e., as $\psi|O_K = 1, \psi|\pi^{-1}O_K \neq 1$. As for $f(x), f^*(x^*)$ we require that each has its coefficients in O_K but not all in πO_K. We observe that after this normalization we can replace $\psi(a), f(x), f^*(x^*)$ only by $\psi(c_0 a), cf(x), c^* f^*(x^*)$ for some c_0, c, c^* in $O_K^\times$ under which $\Gamma(\omega)$ changes as

$$\Gamma(\omega) \to \chi(c_0)^{-d} \chi(cc^*) \cdot \Gamma(\omega), \quad \chi = \omega | O_K^\times.$$

In particular, $\Gamma(\omega_s)$ depends only on the ordering of $Y_1, \ldots, Y_l$.

Now, as in §11, we denote the reduction mod π by $G \to \overline{G}, \ldots$ and make the following assumptions: $\overline{G}$ is connected reductive, acts transitively on $\overline{X} - (\overline{f})^{-1}(0)$, and $\overline{b}_0 \neq 0$. Also, we exclude the trivial case in which $Y(F_q) = \overline{Y}(F_q)$ is empty. Then the correspondence $G(O_K)x \to G(F_q)\overline{x}$ gives a bijection from $G(O_K) \setminus Y(O_K)$ to $G(F_q) \setminus Y(F_q)$, in which $G(F_q) = \overline{G}(F_q)$ and $Y(O_K)$ is the subset of $X(O_K) = O_K^n$ defined by $f(x) \not\equiv 0 \mod \pi$. Furthermore, if we choose the ordering of $Y_1, \ldots, Y_l$ suitably, then we get

$$Y(O_K) = \coprod_{i=1}^{l_0} (Y(O_K) \cap Y_i), \quad Y^*(O_K) = \coprod_{i=1}^{l_0} (Y^*(O_K) \cap Y_i^*)$$

for some $1 \leq l_0 \leq l$. We denote by $\Gamma(\omega)^0$ the square matrix with $\gamma_{ij}(\omega)$ as its (i,j)-entry for $1 \leq i, j \leq l_0$ and restrict ω by the condition that $\omega | (1 + \pi O_K) = 1$. Finally, we regard χ and ψ as characters of $F_q^\times, F_q$ via the group isomorphisms

$$O_K^\times / (1 + \pi O_K) \cong F_q^\times, \quad \pi^{-1} O_K / O_K \cong F_q.$$

Then the following theorem can easily be proved.

THEOREM 14.4. *The submatrix $\Gamma(\omega)^0$ of $\Gamma(\omega)$ has an expression of the form*

$$\Gamma(\omega)^0 = q^{ds} \cdot \sum_{e=0}^{\infty} \Gamma_e(\chi) q^{-es}$$

with $\Gamma_e(\chi)$ independent of s and, if we denote its (i,j)-entry by $\gamma_e(\chi)_{ij}$, then for every a in $Y_j(F_q)$ we have

$$q^{-n} \cdot \sum_{a^* \in Y^*(F_q)} \chi(f^*(a^*)) \psi([a, a^*]) = \chi(a)^{-1} \cdot \sum_{i=1}^{l_0} \gamma_0(\chi)_{ij}.$$

If (G, X) satisfies the condition in Theorem 14.2, then $a_K(G, \omega) = 1$ implies

$$\sum_{e=0}^{\infty} \Gamma_e(\chi) q^{-es} = \chi(b_0) q^{-(n+d)/2} \begin{cases} (-1)^d \prod_\lambda (1 + (1-q) \sum_{e=1}^{\infty} q^{-(s+\lambda-\kappa)e}) \\ (\sum_{c \in F_q^\times} \chi(c) \psi(c))^d \end{cases}$$

respectively for $\chi = 1$ and $\chi \neq 1$. In the general case, by Gyoja's theory in [10] of prehomogeneous vector spaces over finite fields, an explicit form of the right-hand side of the second identity in Theorem 14.4 is known for all large q up to a factor of absolute value 1. As for $\Gamma(\omega)$ itself, it has been computed in the cases in which $f(x)$ is a quadratic form, the discriminant of a binary cubic form, the determinant of a symmetric matrix, etc., mostly for $\omega = \omega_s$.

§15. Conclusion

Finally, we shall summarize those problems that are directly related to the topics covered in this article. We start by repeatedly emphasizing the problem of clarifying the relation between the poles of $|f|_K^s$ and the roots of $b_f(s)$. We recall that $b_f(s)$ is known for the relative invariant of every IRPVS (G, X) but not by a formula. We remark that G is determined by the Dynkin diagram of $D(G)$ and the highest weight as an irreducibly represented group. The problem is to find a closed form of $b_f(s)$ by these data and not only to verify the formula case-by-case but to prove it directly. A similar problem can be proposed for $Z(s)$, namely, to discover a conjectural closed form of $Z(s)$ in terms of the above data and to prove it. This problem is urgent because $Z(s)$ is not yet known for 5 types among the 29 types of IRPVS. The closed form of $Z(s_0, \Phi)$ in Theorem 12.1 contains the operation of taking an average over the Weyl group not only once but twice. There is a strong possibility that the unknown closed form of $Z(s)$ has some similarity. If G_0 is a connected irreducible K-split semisimple algebraic subgroup of GL_n such that its ring of invariants is generated by one polynomial $f(x)$, e.g., $G_0 = D(G)$ in the above notation, we can ask the same question for such an $f(x)$. We might even ask a bigger question for a general $f(x)$ to describe the corresponding $b_f(s)$ and $Z(s)$ in terms of the "invariants of the critical set C_f".

We shall conclude this article by relating Weil's conditions (A), (B), from which we started, to the b-function. Firstly, the following theorem can be derived from Theorem 4.1 by using a remark in [20, pp. 238–239].

THEOREM 15.1. *If $f : K^n \to K$ has no critical value in $K^\times$ and $b_f(s)$ has the form*

$$b_f(s) = (s+1) \cdot \prod_{\lambda > 1}(s + \lambda),$$

then f satisfies Condition (A).

As for Condition (B), the following theorem is proved in [14].

THEOREM 15.2. *Let $f(x)$ denote a homogeneous polynomial in n variables with coefficients in an algebraic number field k and assume that (i) the codimension of C_f in $f^{-1}(0)$ is at least two and (ii) there exists a real number $\sigma > 2$ such that*

$$\left| \int_{O_K^n} \psi(i^* f(x))\, dx \right| \leq \max(1, |i^*|_K)^{-\sigma}$$

for almost all p-adic completions K of k and for all i^* in K, in which ψ is a character of K satisfying $\psi|O_K = 1, \psi|\pi^{-1}O_K \neq 1$. Then $f : A_k^n \to A_k$, where A_k denotes the adèle ring of k, satisfies Condition (B); hence the Poisson formula (P) holds.

If $f(x)$ is a nondegenerate quadratic form, the conditions (i), (ii) clearly become $n \geq 3, n > 4$ while $b_f(s) = (s+1)(s + n/2)$. The following conjecture is a variant of a conjecture in [14].

CONJECTURE 15.3. *Condition (ii) in Theorem 15.2 can be replaced by $b_f(s)$ having the form*

$$b_f(s) = (s+1) \cdot \prod_{\lambda > 2}(s + \lambda).$$

We have restricted our topics by the imposed limitation on the length of the article. We would like to refer the readers to Denef [9] for further theorems, problems, and references.

References

1. M. F. Atiyah, *Resolution of singularities and division of distributions*, Comm. Pure Appl. Math. **23** (1970), 145–150.
2. I. N. Bernstein and S. I. Gel'fand, *Meromorphic property of the functions P^λ*, Functional Anal. Appl. **3** (1969), 68–69.
3. I. N. Bernstein, *The analytic continuation of generalized functions with respect to a parameter*, Functional Anal. Appl. **6** (1972), 273–285.
4. S. I. Borewicz and I. R. Šafarevič, *Zahlentheorie*, Birkhäuser, 1966.
5. J. Denef, *The rationality of the Poincaré series associated to the p-adic points on a variety*, Invent. Math. **77** (1984), 1–23.
6. _____, *On the degree of Igusa's local zeta function*, Amer. J. Math. **109** (1987), 991–1008.
7. J. Denef and L. van den Dries, *p-adic and real subanalytic sets*, Ann. Math. **128** (1988), 79–138.
8. J. Denef and D. Meuser, *A functional equation of Igusa's local zeta function*, Amer. J. Math. **113** (1991), 1135–1152.
9. J. Denef, *Report on Igusa's local zeta function*, Sém. Bourbaki **741** (1991), 1–25.
10. A. Gyoja, *Gauss sums of prehomogeneous vector spaces*, preprint.
11. _____, *Functional equation for Igusa local zeta functions*, JAMI Lecture, 1993.
12. H. Hironaka, *Resolution of singularities of an algebraic variety over a field of characteristic zero*. I–II, Ann. Math. **79** (1964), 109–326.
13. J. Igusa, *Complex powers and asymptotic expansions*. I, Crelles J. Math. **268/269** (1974), 110–130; II, ibid. **278/279** (1975), 307–321.
14. _____, *Criteria for the validity of a certain Poisson formula*, Algebraic Number Theory, Japan Soc. Prom. Sci., 1977, pp. 43–65.
15. _____, *Exponential sums associated with a Freudenthal quartic*, J. Fac. Sci. Univ. Tokyo **24** (1977), 231–246.
16. _____, *Some results on p-adic complex powers*, Amer. J. Math. **106** (1984), 1013–1032.
17. _____, *Complex powers of irreducible algebroid curves*, Geometry Today, Progress in Math., vol. 60, Birkhäuser, 1985, pp. 207–230.
18. _____, *On functional equations of complex powers*, Invent. Math. **85** (1986), 1–29.
19. _____, *On the arithmetic of a singular invariant*, Amer. J. Math. **110** (1988), 197–233.
20. _____, *B-functions and p-adic integrals*, Algebraic Analysis I, Academic Press, 1988, pp. 231–241.
21. _____, *Universal p-adic zeta functions and their functional equations*, Amer. J. Math. **111** (1989), 671–716.
22. _____, *A stationary phase formula for p-adic integrals and its applications*, Algebraic Geometry and its Applications, Springer-Verlag, 1994, pp. 175–194.
23. _____, *Local zeta functions of certain prehomogeneous vector spaces*, Amer. J. Math. **114** (1992), 251–296.
24. _____, *Local zeta functions of general quadratic polynomials*, Proc. Indian Acad. Sci. **104** (1994), 177–189.
25. N. Iwahori and H. Matsumoto, *On some Bruhat decomposition and the structure of the Hecke rings of p-adic Chevalley groups*, Inst. Hautes Études Sci. Publ. Math. **25** (1965), 5–48.
26. M. Kashiwara, *B-functions and holonomic systems (Rationality of b-functions)*, Invent. Math. **38** (1976), 33–53.
27. T. Kimura, *The b-functions and holonomy diagrams of irreducible regular prehomogeneous vector spaces*, Nagoya Math. J. **85** (1982), 1–80.
28. T. Kimura, F. Sato, and X.-W. Zhu, *On the poles of p-adic complex powers and the b-functions of prehomogeneous vector spaces*, Amer. J. Math. **112** (1990), 423–437.
29. F. Loeser, *Fonctions d'Igusa p-adiques et polynômes de Bernstein*, Amer. J. Math. **110** (1988), 1–22.
30. _____, *Fonctions d'Igusa p-adiques, polynômes de Bernstein et polyèdres de Newton*, Crelles J. Math. **412** (1990), 75–96.
31. I. G. MacDonald, *Spherical functions on a group of p-adic type*, Publ. Ramanujan Inst. **2** (1971).
32. A. Macintyre, *On definable subsets of p-adic fields*, J. Symbolic Logic **41** (1976), 605–610.

33. J. Oesterlé, *Réduction modulo p^n des sousensembles analytiques fermés de Z_p^N*, Invent. Math. **66** (1982), 325–341.
34. C.-L. Pan, *A generalization of Tate's local zeta functional equations to certain twisted Igusa local zeta functions*, Thesis, Johns Hopkins, 1993.
35. I. Satake, *Theory of spherical functions on reductive algebraic groups over p-adic fields*, Inst. Hautes Études Sci. Publ. Math. **18** (1964), 5–69.
36. M. Sato, *Theory of prehomogeneous vector spaces* (notes by T. Shintani), Sugaku-no-ayumi **15-1** (1970), 85–156. (Japanese)
37. M. Sato, M. Kashiwara, T. Kimura, and T. Oshima, *Micro-local analysis of prehomogeneous vector spaces*, Invent. Math. **62** (1980), 117–179.
38. M. Sato and T. Kimura, *A classification of irreducible prehomogeneous vector spaces and their relative invariants*, Nagoya Math. J. **65** (1977), 1–155.
39. J.-P. Serre, *Cohomologie Galoisienne*, Lecture Notes in Math., vol. 5, Springer-Verlag, 1965.
40. _____, *Quelques applications du théorème de densité de Chebotarev*, Inst. Hautes Études Sci. Publ. Math. **54** (1981), 123–201.
41. T. Shintani, *On Dirichlet series whose coefficients are class numbers of integral binary cubic forms*, J. Math. Soc. Japan **24** (1972), 132–188.
42. L. Strauss, *Poles of a two-variable p-adic complex power*, Trans. Amer. Math. Soc. **278** (1983), 481–493.
43. A. Weil, *Sur la formule de Siegel dans la théorie des groupes classiques*, Acta Math. **113** (1965), 1–87; Collected Papers III, 71–157.
44. T. Yano and I. Ozeki, *Micro-local structure of the regular prehomogeneous vector space associated with* $SL(5) \times GL(4)$, preprint.

THE JOHNS HOPKINS UNIVERSITY, BALTIMORE, MD

Translated by JUN-ICHI IGUSA

Mahler Functions and Transcendental Numbers

Kumiko Nishioka

§1. Introduction

A complex number α is called algebraic if it is algebraic over the field $\mathbf{Q}$ of rational numbers. A complex number that is not algebraic is called transcendental. When m complex numbers $\alpha_1, \alpha_2, \ldots, \alpha_m$ are algebraically independent over $\mathbf{Q}$, we simply say they are algebraically independent. By a set-theoretical consideration, the set of algebraic numbers is countable, and hence almost all complex numbers are transcendental. It is difficult, however, to know if a given number is transcendental or that a given set of numbers is algebraically independent.

Liouville proved, prior to the birth of Cantor's set theory, the existence of transcendental numbers by actually constructing them. For example, $\sum_{k=1}^{\infty} 10^{-k!}$ is a transcendental number. Later, the base e of the natural logarithm and the number π were shown to be transcendental. Gel'fand and Schneider proved that if α is an algebraic number different from 0 or 1 and β is an algebraic number that is not rational, α^β is transcendental. We have the following theorem concerning values of the exponential function evaluated on algebraic numbers.

THEOREM (Lindemann-Weierstrass). *If $\alpha_1, \ldots, \alpha_m$ are algebraic numbers that are linearly independent over $\mathbf{Q}$, then $e^{\alpha_1}, \ldots, e^{\alpha_m}$ are algebraically independent.*

We have as a generalization of this theorem the Siegel and Shidlovskii theory on E-functions. Let K be an algebraic field and $f(z) = \sum_{n=0}^{\infty} \dfrac{a_n}{n!} z^n$ a power series whose coefficients are in K. We call $f(z)$ an E-function if for some positive constant c, the absolute values of conjugates of a_n are less than or equal to c^{n+1} and $d_n \leq c^{n+1}$, where d_n is the smallest natural number n making each one of $da_0, \ldots, da_n$ an algebraic integer. Suppose that m E-functions $f_1(z), \ldots, f_m(z)$ satisfy differential equations

$$f_i' = \sum_{j=1}^{m} a_{ij} f_j + b_i, \qquad i = 1, \ldots, m,$$

where a_{ij}, b_i are elements of the field $K(z)$ of rational functions over K. Write $T(z)$ for the common denominator of a_{ij}, b_j ($1 \leq i, j \leq m$).

1991 *Mathematics Subject Classification.* Primary 11J81, 11J91.
This article originally appeared in Japanese in Sūgaku **44** (1) (1992), 125–132.

THEOREM (Siegel-Shidlovskii). *If an algebraic number α satisfies $\alpha T(\alpha) \neq 0$ then*

$$\text{trans.deg}_{\mathbf{Q}} \mathbf{Q}(f_1(\alpha),\ldots,f_m(\alpha)) = \text{trans.deg}_{K(z)} K(z)(f_1,\ldots,f_m),$$

where $\text{trans.deg}_{\mathbf{Q}}$ *and* $\text{trans.deg}_{K(z)}$ *are transcendental degrees over $\mathbf{Q}$ and $K(z)$ respectively.*

Liouville's example of a transcendental number is the value of the power series $\sum_{k=1}^{\infty} z^{k!}$ at $\frac{1}{10}$. One can prove that $\sum_{k=1}^{\infty} \alpha^{k!}$ is transcendental for an algebraic number α, $0 < |\alpha| < 1$. A conjecture with regard to the algebraic independency of these values had been that $\sum_{k=1}^{\infty} \alpha_1^{k!},\ldots, \sum_{k=1}^{\infty} \alpha_m^{k!}$ are algebraically independent if and only if, for each pair of distinct i and j, α_i/α_j is not an nth root of unity. The following is a recent result which includes the solution of this conjecture.

THEOREM (Nishioka [32]). *Suppose that the power series $f(z) = \sum_{k=0}^{\infty} z^{e_k}$, $0 \leq e_0 < e_1 < \cdots$, satisfies $\lim_{k\to\infty} e_k/e_{k+1} = 0$. Let $\alpha_1,\ldots,\alpha_m$ be algebraic numbers such that $0 < |\alpha_i| < 1$, $i = 1,\ldots,m$. Then the following conditions are equivalent.*

(i) $f^{(l)}(\alpha_i)$ $(1 \leq i \leq m, l \geq 0)$ *are algebraically dependent ($f^{(l)}(z)$ is the lth derivative of $f(z)$).*

(ii) $1, f(\alpha_1),\ldots,f(\alpha_m)$ *are linearly dependent over $\bar{\mathbf{Q}}$ ($\bar{\mathbf{Q}}$ is the set of algebraic numbers).*

(iii) *There exist a nonempty subset $\{i_1,\ldots,i_n\}$ of $\{1,\ldots,m\}$, an algebraic number γ, the nth roots of unity, $\zeta_1,\ldots,\zeta_n$, and algebraic numbers, $d_1,\ldots,d_n$, not all zero, such that*

$$\alpha_{i_j} = \zeta_j \gamma, \quad j = 1,\ldots,n,$$

and

$$\sum_{j=1}^{n} d_j \zeta_j^{e_k} = 0.$$

Thus if $\alpha_1,\ldots,\alpha_m$, $0 < |\alpha_i| < 1$, $i = 1,\ldots,m$, are algebraic numbers such that if for $1 \leq i < j \leq m$, α_i/α_j is not an nth root of unity, then $f^{(l)}(\alpha_i)$, $1 \leq i \leq m$, $l \geq 0$, are algebraically independent. Further, if we set $f(z) = \sum_{k=1}^{\infty} z^{k!+k}$ then for pairwise distinct $\alpha_1,\ldots,\alpha_m$, $f^{(l)}(\alpha_i)$, $1 \leq i \leq m$, $l \geq 0$, are algebraically independent.

Among other types of functions whose algebraic independence is investigated we find Mahler functions. The main purpose of this paper is to explain this feature of Mahler functions, which we define in the next section.

§2. Mahler functions

Let K be an algebraic field. Let $f_1(z),\ldots,f_m(z)$ be power series with coefficients in K that converge in $|z| < r$. Suppose for some natural number d greater than or equal to two $f_1(z),\ldots,f_m(z)$ satisfy the following functional equations:

$$f_i(z) = \sum_{j=1}^{m} a_{ij}(z) f_j(z^d) + b_i(z), \quad i = 1,\ldots,m,$$

where $a_{ij}(z)$, $b_i(z)$ are elements of $K(z)$. These functions are called Mahler functions. The simplest example of a Mahler function is given by $f(z) =$

$\sum_{k=0}^{\infty} z^{d^k}$ and it satisfies $f(z) = f(z^d) + z$. Denote by $T(z)$ the common denominator of $a_{ij}(z)$, $b_i(z)$. In what follows we assume that α is an algebraic number such that
$$0 < |\alpha| < \min(1, r),$$
$$T(\alpha^{d^k}) \det(a_{ij}(\alpha^{d^k})) \neq 0, \quad k \geq 0.$$

Mahler [19] showed in the case the matrix $(a_{ij}(z))$ is diagonal with constant entries, that if $f_1(z),\ldots,f_m(z)$ are algebraically independent over $K(z)$ then $f_1(\alpha),\ldots,f_m(\alpha)$ are algebraically independent. From this it follows that $\sum_{k=1}^{\infty} \alpha^{d^k}$ is a transcendental number. Kubota then proved that the same theorem holds as long as $(a_{ij}(z))$ is a diagonal matrix not necessarily of constant entries. It had been difficult to treat $(a_{ij}(z))$ in general using Mahler's method; however, with Nesterenko's theory, which recently emerged in transcendental number theory, we prove the following.

THEOREM 1 (Nishioka [34]). *Let $f_1,\ldots,f_m$, α be as above. Then*
$$\text{trans.deg}_{\mathbb{Q}} \mathbb{Q}(f_1(\alpha),\ldots,f_m(\alpha))$$
$$= \text{trans.deg}_{K(z)} K(z)(f_1,\ldots,f_m).$$

Here is an example. For a natural number n we denote by $e(n)$ the number of times "11" appears in its binary representation. We call $a(n) = (-1)^{e(n)}$ the Rudin-Shapiro sequence. Set
$$f(z) = \sum_{n=0}^{\infty} a(n) z^n.$$

Since $a(0) = 1$, $a(2n) = a(n)$, $a(2n+1) = (-1)^n a(n)$, $f(z)$ and $f(-z)$ satisfy the functional equations
$$\begin{pmatrix} f(z) \\ f(-z) \end{pmatrix} = \begin{pmatrix} 1 & z \\ 1 & -z \end{pmatrix} \begin{pmatrix} f(z^2) \\ f(-z^2) \end{pmatrix}.$$

Further, we showed in [34] that $f(z)$ and $f(-z)$ are algebraically independent over $\mathbb{C}(z)$; hence, for an algebraic number α, $0 < |\alpha| < 1$, $f(\alpha)$ and $f(-\alpha)$ are algebraically independent.

We give another example. Let d be a natural number greater than or equal to two. The natural number n has a unique expression
$$n = n_0 + n_1 d + \cdots + n_r d^r, \quad n_r \neq 0, \quad 0 \leq n_0,\ldots,n_r \leq d-1,$$
which we will denote by $n = (n_r,\ldots,n_0)$ with the agreement $0 = (0)$. For $0 \leq t \leq d-1$, $i \geq 0$, define
$$S(t, i) = \{ n = (n_r,\ldots,n_0) \mid t \text{ appears among } n_r,\ldots,n_0 \text{ at most } i \text{ times} \},$$
and set
$$f_{ti}(z) = \sum_{n \in S(t,i)} z^n.$$

The $\{f_{ti}(z)\}_{i \geq 0}$ satisfy the following functional equations. Set $P_t(z) =$

$(1-z^d)/(1-z) - z^t$. Then

$$f_{t0}(z) = \begin{cases} P_0(z)(1 + f_{00}(z^d)) & (t=0) \\ P_t(z)f_{t0}(z^d) & (t \neq 0) \end{cases}$$

$$f_{t1}(z) = \begin{cases} P_0(z)f_{01}(z^d) + f_{00}(z^d) + 1 & (t=0) \\ P_t(z)f_{t1}(z^d) + z^t f_{t0}(z^d) & (t \neq 0) \end{cases}$$

$$f_{ti}(z) = P_t(z)f_{ti}(z^d) + z^t f_{t,i-1}(z^d) \qquad (i \geq 2).$$

When $d \geq 3$ we can show that the $\{f_{ti}(z)\}_{i \geq 0, 0 \leq t \leq d-1}$ are algebraically independent over $\mathbf{C}(z)$ (see [28]); hence, the $\{f_{ti}(\alpha)\}_{i \geq 0, 0 \leq t \leq d-1}$ are algebraically independent if the algebraic number α satisfies $0 < |\alpha| < 1$ and $\prod_{t=0}^{d-1} P_t(\alpha^{d^k}) \neq 0$, $k \geq 0$.

One finds detailed discussions of the relationship between finite automata and Mahler functions in Loxton [13] as well as in Loxton and van der Poorten [16].

We will give an outline of the proof of Theorem 1 in the next section, where the method of Nesterenko is essential. Nesterenko proves that if α is an algebraic number neither 0 nor 1 and β is an algebraic number of degree d, $d \geq 2$, then

$$\text{trans. deg}_\mathbf{Q} \mathbf{Q}(\alpha^\beta, \dots, \alpha^{\beta^{d-1}}) \geq \left[\frac{d+1}{2}\right]$$

(Nesterenko [26], Diaz [10]). Nesterenko [23] offers a historical passage on this problem in detail.

Now that we know by Theorem 1 that the algebraic independence of $f_1, \dots, f_m$ over $K(z)$ implies that of $f_1(\alpha), \dots, f_m(\alpha)$, we wish to consider the degrees of algebraic independence of these numbers. For a polynomial $Q \in \mathbf{C}[x_1, \dots, x_m]$, let $H(Q)$ denote the maximum of the absolute values of its coefficients and $\deg_{\underline{x}} Q$ the total degree with respect to $x_1, \dots, x_m$. Set

$$\varphi(H, s) = \min_{\substack{0 \neq Q \in \mathbf{Z}[x_1, \dots, x_m] \\ H(Q) \leq H \\ \deg_{\underline{x}} Q \leq s}} |Q(f_1(\alpha), \dots, f_m(\alpha))|,$$

where $s, H \geq 1$. Nesterenko [24] gave the inequality

$$\varphi(H, s) \geq \psi(s) H^{-\gamma s^m}, \quad \text{where } \gamma \text{ is a positive constant,}$$

but did not know what sort of function of s $\psi(s)$ was.

THEOREM 2 (Nishioka [36]). *If $f_1, \dots, f_m$ are algebraically independent over $K(z)$,*

$$\varphi(H, s) \geq \exp(-\gamma s^m (\log H + s^{m+2})).$$

If all a_{ij}, b_i are polynomials,

$$\varphi(H, s) \geq \exp\left(-\gamma s^m (\log H + s^2 \log(s+1))\right).$$

In general, we define $\varphi(H, s)$ for algebraically independent numbers $\omega_1, \dots, \omega_m$ and say that the transcendence type of $\mathbf{Q}(\omega_1, \dots, \omega_m)$ is less than or equal to τ (see [6]) if there exist positive constants γ and τ such that

$$\varphi(H, s) \geq \exp\left(-\gamma(s + \log H)^\tau\right).$$

Here we must have $\tau \geq m + 1$. Further, the transcendence types of $\mathbf{Q}(\pi)$ and $\mathbf{Q}(e)$ are less than or equal to $2 + \varepsilon$ (ε is an arbitrary positive number) and 3 respectively. We show that, in the first case of Theorem 2, the transcendence type of $\mathbf{Q}(f_1(\alpha), \ldots, f_m(\alpha))$ is less than or equal to $2m + 2$ and, in the second case, it is less than or equal to $m + 2 + \varepsilon$ (ε is an arbitrary positive number). In particular, the estimate of the second case is quite good in that it exceeds $m + 1$ by $1 + \varepsilon$. In fact, no example of a field $\mathbf{Q}(\omega_1, \ldots, \omega_m)$ of a finite transcendence type for a given m has been known. To prove Theorem 2 we essentially use an estimate of the orders of the zeroes in the following.

THEOREM 3 (Nishioka [35]). *Let $f_1, \ldots, f_m$ be Mahler functions. If $Q \in \mathbf{C}[z, x_1, \ldots, x_m]$ is a polynomial such that*

$$\deg_z Q \leq M, \quad \deg_x Q \leq N, \quad M \geq N \geq 1,$$

and $Q(z, f_1(z), \ldots, f_m(z)) \neq 0$, we have the inequality

$$\operatorname{ord}_{z=0} Q(z, f_1(z), \ldots, f_m(z)) \leq cMN^m,$$

where $\operatorname{ord}_{z=0}$ is the order at $z = 0$ and c is a positive constant determined by $f_1, \ldots, f_m$.

Nesterenko has a theorem for E-functions similar to Theorem 3 (see [25]), but his estimate on the degree of the algebraic independence of values of E-functions is not as sharp as Theorem 2. Another application of Theorem 3 gives an estimate for the values of a Mahler function evaluated at a transcendental number.

THEOREM 4 (Amou [7]). *Suppose that $f_1, \ldots, f_m$ are algebraically independent over $\mathbf{C}(z)$, and a_{ij}, b_i are polynomials. If a transcendental number ω satisfies $0 < |\omega| < \min\{1, r\}$, we have*

$$\operatorname{trans.deg}_{\mathbf{Q}} \mathbf{Q}(\omega, f_1(\omega) \ldots, f_m(\omega)) \geq \left[\frac{m+1}{2}\right].$$

One would attain the best estimate if one could replace $[(m + 1)/2]$ by m, but it looks difficult. Another limitation is that Nesterenko's method is applicable only to power series of one variable, although Mahler dealt with power series of several variables. These are yet to be solved. We gave the detailed proofs of Theorems 1 through 4 in Nishioka [4].

§3. Outline of the proof of Theorem 1

We assume for brevity that $K = \mathbf{Q}$ and $\alpha \in \mathbf{Q}$. Let n be the transcendental degree of $\mathbf{Q}(z)(f_1, \ldots, f_m)$ over $\mathbf{Q}(z)$. We may assume that $f_1, \ldots, f_n$ are algebraically independent over $\mathbf{Q}(z)$. If $\operatorname{trans.deg}_{\mathbf{Q}} \mathbf{Q}(f_1(\alpha), \ldots, f_m(\alpha)) = n_0$, it is easy to see that $n_0 \leq n$; hence, we only need to show $n_0 \geq n$. Moreover, it is enough to consider $n \geq 1$.

PROPOSITION 1. *Let N be a natural number. Then there exists a polynomial $R(z, x_1, \ldots, x_n) \in \mathbf{Z}[z, x_1, \ldots, x_n]$ satisfying the following:*
  (i) $\deg_z R \leq N$, $\deg_x R \leq N$.
  (ii) $\operatorname{ord}_{z=0} R(z, f_1(z), \ldots, f_n(z)) \geq N^{n+1}/n!$.

PROOF. We regard the coefficients of R as variables and show that the system of $N^{n+1}/n!$ linear equations in $(N+1)\binom{N+n}{n}$ variables has nontrivial solutions.

We set
$$E_N(z) = R(z, f_1(z), \ldots, f_n(z)).$$

Then, since $f_1, \ldots, f_n$ are algebraically independent over $\mathbf{Q}(z)$, $E_N(z)$ is different from zero. Put $\mathrm{ord}_{z=0} E_N = l(N)$ $(\geq N^{n+1}/n!)$. We denote by $c(N)$ a positive constant determined by N. Since $E_N(z)$ is a power series that converges for $|z| < r$, we have for $k \geq c(N)$

$$-c_1 d^k l(N) \leq \log |E_N(\alpha^{d^k})| \leq -c_2 d^k l(N).$$

In what follows, $c_1, c_2, \ldots$ are positive constants depending only on $f_1, \ldots, f_m$ and α. Using the functional equations repeatedly we express $f_1(\alpha^{d^k}), \ldots, f_m(\alpha^{d^k})$ as a linear combination of $1, f_1(\alpha), \ldots, f_m(\alpha)$ over $\mathbf{Q}$. Hence we may write

$$\begin{aligned} E_N(\alpha^{d^k}) &= R(\alpha^{d^k}, f_1(\alpha^{d^k}), \ldots, f_n(\alpha^{d^k})) \\ &= R_k(f_1(\alpha), \ldots, f_m(\alpha)), \end{aligned}$$

where $R_k \in \mathbf{Q}[x_1, \ldots, x_m]$. Multiplying R_k by the denominators of the coefficients we obtain an integer polynomial Q_k that satisfies the following.

PROPOSITION 2. For $k \geq c(N)$,
(i) $\deg_x Q_k \leq N$,
(ii) $\log H(Q_k) \leq c_3 d^k N$,
(iii) $-c_4 d^k l(N) \leq \log |Q_k(f_1(\alpha), \ldots, f_m(\alpha))| \leq -c_5 d^k l(N)$.

We now suppose $n_0 < n$ to get a contradiction. Before discussing the general situation, we first present the elementary case, $n_0 = 0$. Set

$$(f_1(\alpha), \ldots, f_m(\alpha)) = (\omega_1, \ldots, \omega_m) = \underline{\omega},$$

and denote by $\underline{\omega}^{(1)} (= \underline{\omega}), \ldots, \underline{\omega}^{(g)}$ the conjugates of $\underline{\omega}$ in $\mathbf{Q}$. Choose a natural number a such that each one of $a\omega_1, \ldots, a\omega_m$ is an algebraic integer. Then

$$\prod_{i=1}^{g} a^N Q_k(\underline{\omega}^{(i)})$$

turns out to be a rational integer since it is an algebraic integer that is rational as well. Further, by Proposition (iii), if $k \geq c(N)$, we have

$$0 \leq \log |\prod_{i=1}^{g} a^N Q_k(\underline{\omega}^{(i)})| \leq -c_5 d^k l(N) + Ng \log a + c_6 d^k N.$$

Dividing both sides of the inequality by d^k and letting $k \to \infty$, we get $c_5 l(N) \leq c_6 N$, but this is a contradiction for a sufficiently large N since $l(N) \geq N^{n+1}/n!$.

In the general setting, we again choose a suitable nonzero integer under the assumption $n_0 < n$ and at the same time show that its absolute value must be less than 1, which leads to a contradiction. We introduce some notation.

Denote by $\mathscr{P}$ a prime ideal of the polynomial ring $\mathbf{Z}[x_0, \ldots, x_m]$ of $m+1$ variables with $\mathscr{P} \cap \mathbf{Z} = (0)$ and $h(\mathscr{P})$ its height. Assume $1 \leq h(\mathscr{P}) \leq m$ and put $r = m + 1 - h(\mathscr{P})$. We consider r linear forms of the variables u_{ij}, $1 \leq i \leq r$, $0 \leq j \leq m$,

$$L_i(\underline{x}) = \sum_{j=0}^{m} u_{ij} x_j, \quad 1 \leq i \leq r.$$

Let $(\mathscr{P}, L_1, \ldots, L_r)$ be the ideal in $\mathbf{Z}[x_0, \ldots, x_m, u_{10}, \ldots, u_{rm}]$ generated by $\mathscr{P}, L_1, \ldots, L_r$. Let $\bar{\mathscr{P}}(r)$ be the ideal consisting of the elements G of $\mathbf{Z}[u_{10}, \ldots, u_{rm}]$ such that $G x_i^M \in (\mathscr{P}, L_1, \ldots, L_r)$, $0 \leq i \leq m$, for some natural number M. Then $\bar{\mathscr{P}}(r)$ is a nontrivial principal ideal. The generator F is unique up to sign and is homogeneous with respect to $u_{i0}, \ldots, u_{im}$ and its degree does not depend on i. We set $N(\mathscr{P}) = \deg_{u_1} F$. Denote by $H(\mathscr{P})$ the maximum $H(F)$ among the absolute values of the coefficients of F. We call F the Chow form of $\mathscr{P}$. Consider variables $s_{jk}^{(i)}$, $1 \leq i \leq r$, $0 \leq j < k \leq m$, and define a skew-symmetric matrix $S^{(i)} = (s_{jk}^{(i)})$ by setting $s_{jk}^{(i)} + s_{kj}^{(i)} = 0$. Define a ring homomorphism θ from $\mathbf{Z}[u_{10}, \ldots, u_{rm}]$ to $\mathbf{Z}[x_0, \ldots, x_m][\{s_{jk}^{(i)}\}_{1 \leq i \leq r,\, 0 \leq j < k \leq m}]$ by

$$\begin{pmatrix} \theta(u_{i0}) \\ \vdots \\ \theta(u_{im}) \end{pmatrix} = S^{(i)} \begin{pmatrix} x_0 \\ \vdots \\ x_m \end{pmatrix}.$$

Then we see that the coefficients of $\theta(F)$ belong to $\mathscr{P}$ ($\subset \mathbf{Z}[x_0, \ldots, x_m]$). Set

$$(1, f_1(\alpha), \ldots, f_m(\alpha)) = (\omega_0, \omega_1, \ldots, \omega_m) = \underline{\omega},$$

and for $E \in \mathbf{Z}[u_{10}, \ldots, u_{rm}]$ set

$$\kappa(E) = \theta(E)|_{x_i = \omega_i,\, i=0,\ldots,m} \in \mathbf{C}\left[\{s_{jk}^{(i)}\}_{1 \leq i \leq r,\, 0 \leq j < k \leq m}\right].$$

Further, define

$$|\mathscr{P}(\underline{\omega})| = H(\kappa(E)) (\max_{0 \leq i \leq m} |\omega_i|)^{-rN(\mathscr{P})};$$

i.e., think of $\theta(F)$ as a polynomial in variables $\{s_{jk}^{(i)}\}$ and substitute $\underline{\omega}$ in its coefficients (these belong to $\mathscr{P}$). Then $|\mathscr{P}(\underline{\omega})|$ is approximately equal to the maximum of the absolute values of $\theta(F)$ after the substitution.

Now let $\mathscr{P}^{(0)} \subset \mathbf{Z}[x_0, \ldots, x_m]$ be the principal ideal consisting of the homogeneous polynomials that take on the value zero at $\underline{\omega}$. We have $h(\mathscr{P}^{(0)}) = m - n_0$ and $|\mathscr{P}^{(0)}(\underline{\omega})| = 0$. The Chow form F_0 of $\mathscr{P}^{(0)}$ has an expression

$$F_0 = a \prod_{j=1}^{N(\mathscr{P}^{(0)})} \left(u_{r0} + \alpha_1^{(j)} u_{r1} + \cdots + \alpha_m^{(j)} u_{rm}\right),$$

where $r = n_0 + 1$, $a \in \mathbf{Z}[\underline{u}_1, \ldots, \underline{u}_{r-1}]$, $\underline{u}_i = (u_{i0}, \ldots, u_{im})$, and $(1, \alpha_1^{(1)}, \ldots, \alpha_m^{(1)})$ is algebraic over $\mathbf{Z}[\underline{u}_1, \ldots, \underline{u}_{r-1}]$, whose conjugates are given by $\{(1, \alpha_1^{(j)}, \ldots, \alpha_m^{(j)})\}_{1 \leq j \leq N(\mathscr{P}^{(0)})}$; these also generate the zeros of $\mathscr{P}^{(0)}$ over $\mathbf{Q}$. Let N

be a sufficiently large natural number and let k_1 also be a sufficiently large natural number but dependent on N. Using Q_{k_1} of Proposition 2 set

$$G_1 = a^{\deg_{\underline{x}} Q_{k_1}} \prod_{j=1}^{N(\mathscr{P}^{(0)})} Q_{k_1}\left(\alpha_1^{(j)}, \ldots, \alpha_m^{(j)}\right).$$

Then we have $0 \neq G_1 \in \mathbf{Z}[\underline{u}_1, \ldots, \underline{u}_{r-1}]$, and we have the following estimates:

$$\deg_{\underline{u}_1} G_1 \leq c_7 N(\mathscr{P}^{(0)}) N \leq c_8 N,$$

$$\log H(G_1) \leq c_7 \left\{\log H(\mathscr{P}^{(0)}) + d^{k_1} N(\mathscr{P}^{(0)})\right\} N \leq c_8 d^{k_1} N,$$

$$\log H(\kappa(G_1)) \leq -c_9 d^{k_1} l(N) + c_{10} \left\{\log H(\mathscr{P}^{(0)}) + d^{k_1} N(\mathscr{P}^{(0)})\right\} N$$

$$\leq -c_{11} d^{k_1} l(N).$$

Here if $G_1 = b E_1^{r_1} \cdots E_t^{r_t}$ (b is an integer and E_i is an irreducible polynomial) is a prime factorization, each E_i is the Chow form of some prime ideal $\mathscr{P}_i^{(1)}$ ($h(\mathscr{P}_i^{(1)}) = m - n_0 + 1$). The above inequalities show that $\log H(\kappa(G_1))$ is small in comparison with $\deg_{\underline{u}_1} G_1$ or $\log H(G_1)$. This must also be true for some prime factor E_i of G_1, i.e., there exists an i making

$$\log |\mathscr{P}_i^{(1)}(\underline{\omega})| \leq -c_{12} \left\{\log H(\mathscr{P}_i^{(1)}) + d^{k_1} N(\mathscr{P}_i^{(1)})\right\} l(N)/N.$$

We use this Chow form F_1 ($= E_i$) of $\mathscr{P}_i^{(1)}$ and Q_{k_2} of Proposition 2 for a suitable $k_2 \leq k_1$ to construct G_2 as before satisfying $0 \neq G_2 \in \mathbf{Z}[\underline{u}_1, \ldots, \underline{u}_{r-2}]$. Further, we get

$$\deg_{\underline{u}_1} G_2 \leq c_{13} N(\mathscr{P}_i^{(1)}) N,$$

$$\log H(G_2) \leq c_{13} \left\{\log H(\mathscr{P}_i^{(1)}) + d^{k_1} N(\mathscr{P}_i^{(1)})\right\} N,$$

$$\log H(\kappa(G_2))$$

$$\leq -c_{14} \left\{\log H(\mathscr{P}_i^{(1)}) + d^{k_1} N(\mathscr{P}_i^{(1)})\right\} l(N)/N$$

$$+ c_{15} \left\{\log H(\mathscr{P}_i^{(1)}) + d^{k_1} N(\mathscr{P}_i^{(1)})\right\} N$$

$$\leq -c_{16} \left\{\log H(\mathscr{P}_i^{(1)}) + d^{k_1} N(\mathscr{P}_i^{(1)})\right\} l(N)/N.$$

Notice that during this process one can move k_2 around by Proposition 2 (iii) to obtain a desirable value of $|Q_{k_2}(\underline{\omega})|$. We decompose this G_2 into prime factors and obtain as before a prime ideal $\mathscr{P}^{(2)}$, $h(\mathscr{P}^{(2)}) = m - n_0 + 2$, which satisfies the inequality

$$\log |\mathscr{P}^{(2)}(\underline{\omega})| \leq -c_{17} \left\{\log H(\mathscr{P}^{(2)}) + d^{k_1} N(\mathscr{P}^{(2)})\right\} l(N)/N^2.$$

We repeat this operation n_0 times to get a prime ideal $\mathscr{P}^{(n_0)}$, $h(\mathscr{P}^{(n_0)}) = m$, with the inequality

$$\log |\mathscr{P}^{(n_0)}(\underline{\omega})| \leq -c_{18} \left\{\log H(\mathscr{P}^{(n_0)}) + d^{k_1} N(\mathscr{P}^{(n_0)})\right\} l(N)/N^{n_0}.$$

As before, using the Chow form $F_{n_0} \in \mathbf{Z}[\underline{u}_1]$ of $\mathscr{P}^{(n_0)}$ and $Q_{k_{n_0+1}}$ of Proposition 2 for some $k_{n_0+1}, k_{n_0+1} \leq k_1$, we may construct G_{n_0+1}. This time we have $0 \neq G_{n_0+1} \in \mathbf{Z}$. We also have

$$\log H(\kappa(G_{n_0+1}))$$
$$\leq -c_{19}\left\{\log H(\mathscr{P}^{(n_0)}) + d^{k_1}N(\mathscr{P}^{(n_0)})\right\}l(N)/N^{n_0}$$
$$+ c_{20}\left\{\log H(\mathscr{P}^{(n_0)}) + d^{k_1}N(\mathscr{P}^{(n_0)})\right\}N.$$

The L.H.S. of this inequality is $\log|G_{n_0+1}|$ which is nonnegative. By our assumption, $n > n_0$, we have

$$l(N)/N^{n_0} \leq N^{n+1-n_0}/n! \leq N^2/n!.$$

Hence, for a large enough N the R.H.S. becomes negative—a contradiction. This establishes Theorem 1.

The proofs of Theorems 2 and 4 are similar. We prove Theorem 3 using $\mathbf{C}[z]$ in place of $\mathbf{Z}$ and the Chow form of some ideal in $\mathbf{C}[z][x_0, \ldots, x_m]$.

References

Books.
1. A. Baker, *Transcendental Number Theory*, Cambridge Univ. Press, 1975.
2. K. Mahler, *Lectures on Transcendental Numbers*, Lecture Notes in Math., vol. 546, Springer-Verlag, 1976.
3. T. Mitsui, *Analytic Number Theory*, Kyoritsu, 1977. (Japanese)
4. K. Nishioka, *Mahler Functions and Transcendental Numbers*, Seminar on Math. Sci., No. 17, Keio Univ., 1991. (Japanese)
5. A. B. Shidlovskii, *Transcendental Numbers*, Walter de Gruiyter, 1989.
6. M. Waldschmidt, *Nombres Transcendants*, Lecture Notes in Math., vol. 402, Springer-Verlag, 1974.

Papers.
7. M. Amou, *Algebraic independence of the values of certain functions at a transcendental number*, Acta Arith. (to appear).
8. P. Becker and G. Landeck, *Transcendence measure by Mahler's transcendence method*, Bull. Austral. Math. Soc. **33** (1986), 59–65.
9. _____, *Effective measures for algebraic independence of the values of Mahler type functions*, Acta Arith. (to appear).
10. G. Diaz, *Grands degrés de transcendence pour les familles d'exponentielle*, J. Number Theory **31** (1989), 1–23.
11. A. O. Galochkin, *Transcendence measure of functions satisfying certain functional equations*, Math. Notes **27** (1980), 83–88.
12. K. K. Kubota, *On the algebraic independence of holomorphic solutions of certain functional equations and their values*, Math. Ann. **227** (1977), 9–50.
13. J. H. Loxton, *Automata and transcendence*, New Advances in Transcendence Theory (A. Baker, ed.), Cambridge Univ. Press, 1988, pp. 215–228.
14. J. H. Loxton and A. J. van der Pooreten, *Arithmetic properties of certain functions in several variables* II, J. Austral. Math. Soc. Ser. A **24** (1977), 393–408.
15. _____, *Algebraic independence properties of the Fredholm series*, Austral. Math. Soc. Ser. A **26** (1978), 31–45.
16. _____, *Arithmetic properties of automata, Regular sequences*, J. reine angew. Math. **392** (1988), 57–69.
17. K. Mahler, *Arithmetische Eigenschaften der Lösungen einer Klasse von Funktionalgleichungen*, Math. Ann. **101** (1929), 342–366.
18. _____, *Uber das Verschwinden von Verlánderlichen in speziellen Punktfolgen*, Math. Ann. **103** (1930), 573–587.
19. _____, *Arithmetische Eigenschaften einer Klasse transzental-transzendenter Funktionen*, Math. Z. **32** (1930), 545–585.
20. W. M. Miller, *Transcendence measures by a method of Mahler*, J. Austral. Math. Soc. Series A **32** (1982), 68–78.

21. Yu. V. Nesterenko, *Estimates for the orders of zeros of functions of a certain class and applications in the theory of transcendental numbers*, Izv. Akad. Nauk SSSR Ser. Mat. **41** (1977), 253–284; English transl. in Math. USSR Izv. **11** (1977), 239–270.
22. _____, *Estimates for the characteristic function of a prime ideal*, Mat. Sb. **123** (165) (1984), 11–34; English transl. in Math. USSR Sb. **51** (1985), 9–32.
23. _____, *On algebraic independence of algebraic powers of algebraic numbers*, Mat. Sb. **123** (165) (1984), 435–459; English transl. in Math. USSR Sb. **51** (1985), 429–454.
24. _____, *On a measure of the algebraic independence of the values of certain functions*, Mat. Sb. **128** (170) (1985); English transl. in Math. USSR Sb. **56** (1987), 545–567.
25. _____, *On zero estimates of functions of certain classes*, Acta Arith. **53** (1989), 29–46.
26. _____, *Degrees of transcendence of some fields that are generated by values of an exponential function*, Mat. Zametki **46** (1989), no. 3, 40–49; English transl. in Math. Notes **46** (1989), no. 3–4, 706–712.
27. Keiji Nishioka, *Algebraic solutions of certain class of functional equations*, Arch. Math. **44** (1985), 330–335.
28. Keiji Nishioka and Kumiko Nishioka, *Algebraic independence of functions satisfying a certain system of functional equations* (to appear).
29. K. Nishioka, *On a problem of Mahler for transcendency of function values*, J. Austral. Soc. (**Ser. A**) **33** (1988), 386–393.
30. _____, *On a problem of Mahler for transcendency of function values II*, Tsukuba J. Math **7** (1983), 265–279.
31. _____, *Proof of Masser's conjecture on the algebraic independence of values of Liouville series*, Proc. Japan Academy **62** (Ser. A) (1986), 219–222.
32. _____, *Conditions for algebraic independence of certain power series of algebraic numbers*, Comp. Math. **62** (1987), 53–61.
33. _____, *Algebraic independence of certain power series*, Séminaire de Théorie des Nombres de Paris (1987/1988), 201–212.
34. _____, *New approach in Mahler's method*, J. reine angew. Math. **407** (1990), 202–219.
35. _____, *On an estimate for the zeros of Mahler type functions*, Acta Arith. **56** (1990), 249–256.
36. _____, *Algebraic independence measures of the values of Mahler functions*, J. reine angew. Math. (to appear).
37. K. Nishioka and T. Töpfer, *Transcendence measures and nonlinear functional equations of Mahler type*, Arch. Math. (to appear).
38. P. Philippon, *Critères pour l'indépendence algébrique*, Inst. Hautes Études Sci. Publ. Math. **64** (1986), 5–52.
39. N. Ch. Wass, *Algebraic independence of the values of a class of functions considered by Mahler*, Dissertationes Mathematicae **CCCIII** (1990).

MATHEMATICS, HIYOSHI CAMPUS, KEIO UNIVERSITY, 4-1-1 HIYOSHI, YOKOHAMA, 223 JAPAN

Translated by KIKI HUDSON ARAI

Generalization of Class Field Theory

Kazuya Kato

In this article, we mean by an algebraic number field a finite extension field of the rational number field — a so-called finite algebraic number field.

We shall discuss generalization of class field theory for algebraic number fields to the case of finitely generated fields (not only fields of finite degree) over their prime subfields and various new problems and themes that arise in the generalized theory.

One reason why algebraic number fields are studied in number theory is that the rational number field has long been a familiar object to human beings and the study of the rational number field leads naturally to the study of its finite extensions — algebraic number fields; yet another reason is that an algebraic number field is a field that has an especially rich theory and is well worth being the object of our investigation:

algebraic number field $\overset{(1)}{=}$ field that can be an object of number theory

$\overset{(2)}{=}$ field that has an especially rich theory

$\overset{(3)}{=}$ field that is familiar to human beings.

The existence of Equality (2) must be one of the beliefs of number theorists, and the existence of Equality (3) is absolutely doubtless, though it is a wonder and I do not know why. Recent developments in number theory seem to suggest that the left-hand side of Equality (1) should be replaced by

finitely generated field over a prime field $\overset{(1)}{=}$

(for example, zeta functions are considered not only for algebraic number fields but also for finitely generated fields over prime fields).

Class field theory is a theory on abelian extensions of algebraic number fields. Various dreams come to mind if we assume it is possible to generalize it to a theory on abelian extensions of finitely generated fields over prime fields. For example, the function field of a variety $\sum_{i=0}^{m} a_i T_i^n = 0$ $(a_i \in F^\times)$ of Fermat type over an algebraic number field F can be regarded naturally as an abelian extension, with Galois group $(\mathbb{Z}/n\mathbb{Z})^m$, of the rational function field in $m - 1$ variables over F, if F contains the primitive nth roots of unity (cf. (4.2.6)). I am sorry to relate my

---

1991 *Mathematics Subject Classification*. Primary 11R37, 11G45, 14F20.

This article originally appeared in Japanese in Sūgaku **40** (4) (1988), 289–311.

dreams from the beginning, but how wonderful it would be if our generalized class field theory could play a significant role in the arithmetic of such algebraic varieties! Those results and problems concerning class field theory that one finds in books on number theory should generalize to such fields that are finitely generated over prime fields. Furthermore, besides generalization of facts on algebraic number fields, new interesting problems will arise that did not exist in the case of algebraic number fields. These problems include considering, to cite a few, for each abelian extension field, what kind of singularities appear in a model of the field over the integer ring and how far the model is from having good reduction. It is amusing to believe that traditional techniques and viewpoints of classical class field theory, with appropriate formulation, should be applicable to such problems. I would like to believe that Hilbert's words in his *Zahlbericht* that in the theory of abelian extensions "eine Fülle der kostbarsten Schätze noch verborgen liegt" still remain true today.

It is due to suggestions of Professors Yukiyoshi Kawada and Yasutaka Ihara that I aspired to class field theory. (I chose as the theme of my Master Thesis the problem, which was proposed in [6], of constructing class field theory for certain p-adically complete function fields.) I would like to thank them both and also Shuji Saito, with whom I collaborated, and Hiroo Miki, who was a senior fellow in our Department and who influenced me.

We recall in §1 class field theory for algebraic number fields, state its generalization in §2, supplement it in §3, and explain ramification theory in §4.

§1. Class field theory for algebraic number fields

We give in this section a summary of class field theory for algebraic number fields. I tried to keep it readable for non-specialists, to explain how abstract theorems of class field theory imply theorems classically known in number theory, such as the "reciprocity law for quadratic residues", and to emphasize the Principal Spirit of class field theory, which I believe to be something like the following:

(1.0.1) We must be able to tell how abelian extensions of an algebraic number field exist, and what happens in each abelian extension of an algebraic number field, so perfectly as if it were happening on the palms of our hands.

As a content of "know well what happens", we have the existence of a "decomposition law" in abelian extensions, which will be explained in §1.1 below. The facts stated in §1.1 and the reciprocity law of quadratic residues, which has been known since before class field theory, are all "incarnations" of class field theory and can be explained by class field theory (§1.3).

1.1. Examples. In the following, we denote by o_K the integer ring of an algebraic number field K.

EXAMPLE 1.1.1. Let $K = \mathbb{Q}$ and $L = \mathbb{Q}(i)$ with $i = \sqrt{-1}$. Then $o_K = \mathbb{Z}$, $o_L = \mathbb{Z}[i] = \{a + bi;\ a, b \in \mathbb{Z}\}$, and o_L is a unique factorization domain. If p is a prime number, it may not be a prime element in o_L; we consider its decomposition into prime factors in o_L. The following laws hold.

(i) If $p \equiv 1 \bmod 4$, then p is a product $\pi_1 \pi_2$ of two prime elements in o_L, where π_1 and π_2 do not divide each other in o_L (Example: $5 = (2+i)(2-i)$, $13 = (2+3i)(2-3i)$).

(ii) If $p \equiv 3 \bmod 4$, then p remains prime in o_L.

(iii) For $p = 2$, we have $2 = (1+i)^2 \times$ (unit of o_L), and $(1+i)$ is a prime element of o_L.

To summarize, the prime factorization in o_L of a prime number is described by the simple criterion "mod 4".

REMARK. As the reader might have realized from the above equality $5 = (2+i)(2-i) = 2^2 + 1^2$, one deduces the following classical theorem in number theory using (1.1.1): a prime number p can be written in the form $x^2 + y^2$ with some integers x, y if and only if $p \equiv 1 \bmod 4$ or $p = 2$.

EXAMPLE 1.1.2. Let $K = \mathbb{Q}$ and $L = \mathbb{Q}(\sqrt{5})$. Then $o_L = \mathbb{Z}[(1+\sqrt{5})/2]$ and this is also a unique factorization domain. For a prime number p, we have the following.

(i) If $p \equiv 1$ or $4 \bmod 5$, then p is a product $\pi_1 \pi_2$ of two prime elements in o_L, where π_1 and π_2 do not divide each other in o_L.

(ii) If $p \equiv 2$ or $3 \bmod 5$, then p remains prime in o_L.

(iii) For $p = 5$, we have $5 = (\sqrt{5})^2$, and $\sqrt{5}$ is a prime element of o_L.

Let K be an algebraic number field in general and L a finite extension of K. Unique factorization then no longer holds in a general o_K or o_L, though both o_K and o_L were unique factorization domains in Examples (1.1.1) and (1.1.2). Nevertheless, prime ideal decomposition holds in general, and thus talking about prime ideals is more appropriate than talking about prime elements.

In general, let $\mathfrak{p}$ be a nonzero prime ideal of o_K. Suppose the ideal $\mathfrak{p}o_L$ of o_L generated by $\mathfrak{p}$ decomposes into a product of prime ideals

$$\mathfrak{p}o_L = \mathfrak{P}_1^{e_1} \cdots \mathfrak{P}_g^{e_g},$$

where $\mathfrak{P}_1, \ldots, \mathfrak{P}_g$ are distinct prime ideals of o_L. We say that $\mathfrak{p}$ is *unramified* in L/K if $e_1 = \cdots = e_g = 1$, and that $\mathfrak{p}$ is *ramified* in L/K if it is not unramified, i.e., if $e_i \geq 2$ for some $i \geq 1$. There are only finitely many nonzero prime ideals of o_K that are ramified in L/K, and they are determined fairly easily. We have $g \leq [L:K]$ in general; we say that $\mathfrak{p}$ *splits completely* in L/K if $g = [L:K]$ (in this case, $\mathfrak{p}$ is necessarily unramified in L/K). For example, the contents of (1.1.1) and (1.1.2) can be rephrased as follows: if $K = \mathbb{Q}$ and $L = \mathbb{Q}(\sqrt{-1})$, for a prime ideal $p\mathbb{Z}$ (with p a prime number) of $o_K = \mathbb{Z}$, we have

$p\mathbb{Z}$ is ramified in $L/K \Leftrightarrow p = 2$,

$p\mathbb{Z}$ splits completely in $L/K \Leftrightarrow p \equiv 1 \bmod 4$.

If $K = \mathbb{Q}$ and $L = \mathbb{Q}(\sqrt{5})$, we have

$p\mathbb{Z}$ is ramified in $L/K \Leftrightarrow p = 5$,

$p\mathbb{Z}$ splits completely in $L/K \Leftrightarrow p \equiv 1$ or $4 \bmod 5$.

According to class field theory, there exists such a simple decomposition law as in these examples for each finite abelian extension L/K of algebraic number fields.

EXAMPLE 1.1.3. Let $K = \mathbb{Q}$ and $L = \mathbb{Q}(\zeta_n)$ with ζ_n an nth primitive root of unity. If n is even, assume n is divisible by 4. Then for a prime number p, we have

$p\mathbb{Z}$ is ramified in $L/K \Leftrightarrow p|n$,

$p\mathbb{Z}$ splits completely in $L/K \Leftrightarrow p \equiv 1 \bmod n$.

(This implies (1.1.1), which is the case $n = 4$.)

1.2. Main theorem of class field theory. The main theorem of class field theory assumes an abstract form which seems, at first sight, to have nothing to do with what was stated in §1.1.

For any field k, we denote by k^{ab} a maximal abelian extension, that is, the composite field of all finite abelian extensions of k in a fixed algebraic closure of k. To know how abelian extensions exist, it is enough to know the Galois group $\mathrm{Gal}(k^{\mathrm{ab}}/k)$. For, by Galois theory, the finite extensions of k contained in k^{ab} are in one-to-one correspondence with the open subgroups of $\mathrm{Gal}(k^{\mathrm{ab}}/k)$, the correspondence being of such an extension field to its fixing subgroup of $\mathrm{Gal}(k^{\mathrm{ab}}/k)$.

The above is valid for any field k, whereas the purport of class field theory is:

(1.2.1) For special fields k, the Galois group $\mathrm{Gal}(k^{\mathrm{ab}}/k)$ can be approximated by another group which has an explicit description; to know this group is to know $\mathrm{Gal}(k^{\mathrm{ab}}/k)$ and thus abelian extensions of k.

Algebraic number fields are such special fields; local fields of an algebraic number field are also special fields in this sense (§1.4); and there are other special fields, as will be explained in §2 on generalizations.

For an algebraic number field K, the group that approximates $\mathrm{Gal}(K^{\mathrm{ab}}/K)$ is the idele class group C_K of K (the definition of C_K will be given later).

MAIN THEOREM OF CLASS FIELD THEORY (1.2.2). *If K is an algebraic number field, there exists a canonical homomorphism*

$$C_K \to \mathrm{Gal}(K^{\mathrm{ab}}/K)$$

that is approximately an isomorphism.

By saying "approximately an isomorphism", I mean the following: first, the homomorphism is surjective, and its kernel coincides with the connected component (with respect to the natural topology) of the unit element. Second, the homomorphism establishes a one-to-one correspondence between the open subgroups of the two groups. (Hence, the finite abelian extensions are in one-to-one correspondence with the open subgroups of C_K.)

This theorem tells us "how abelian extensions exist", which is part of the Principal Spirit of class field theory (1.0.1). As to "what happens in each finite abelian extension" — for example, to know, for each nonzero prime ideal $\mathfrak{p}$ of o_K, how the ideal $\mathfrak{p} o_L$ is decomposed into prime ideals in o_L — we need the Complementary Theorem (1.2.5) below.

Before that, we give the definition of the idele class group of an algebraic number field. First we need:

DEFINITION 1.2.3. Let K be an algebraic number field. A *place* of K is an embedding (considered modulo isomorphisms) of K, with dense image, into a locally compact nondiscrete field.

For example, if $K = \mathbb{Q}$, the places of $\mathbb{Q}$ are the (usual) embedding of $\mathbb{Q}$ into $\mathbb{R}$ and the embeddings of $\mathbb{Q}$ into the p-adic fields $\mathbb{Q}_p$ defined for each prime number p. The places of an algebraic number field K consist of at most a finite number of embeddings of K into $\mathbb{R}$ — called *real places*; at most a finite number of embeddings of K into $\mathbb{C}$ (two embeddings exchanged by the complex conjugation are regarded as the same place) with dense image (or what amounts to the same, with image not contained in $\mathbb{R}$) — called *complex places*; and infinitely many embeddings of K into the p-adic fields $K_\mathfrak{p}$, which are defined for each nonzero prime ideal $\mathfrak{p}$ of o_K — called *finite places*. Here $K_\mathfrak{p}$ is the fraction field of the $\mathfrak{p}$-adic completion $\hat{o}_{K,\mathfrak{p}}$ of o_K. (Real places and complex places are called *infinite places*.) In the following, we identify a finite place v with the corresponding nonzero prime ideal of o_K. For a place v of K, we write K_v for the locally compact field into which K is embedded by v, and call it the *local field* of K at v.

DEFINITION 1.2.4. The *idele group* $\underline{\prod}_v K_v^\times$ of an algebraic number field K is a subgroup of the direct product $\prod_v K_v^\times$ (where v runs through all places of K) defined below, and the *idele class group* of K is defined as

$$C_K = (\underline{\prod}_v K_v^\times)/K^\times,$$

where $K^\times$ is embedded into $\underline{\prod}_v K_v^\times$ diagonally.

The definition of $\underline{\prod}_v K_v^\times$, which is somewhat complicated, is as follows: for each finite place v, let $V_v \subset K_v^\times$ be the multiplicative subgroup consisting of the invertible elements of $\hat{o}_{K,v}$. Then define

$$\underline{\prod}_v K_v^\times = \{(a_v)_v \in \prod_v K_v^\times; \ a_v \in V_v \text{ for almost all finite places } v\}.$$

We omit explanation of the topology defined on $\underline{\prod}_v K_v^\times$ and C_K.

We have the following Complementary Theorem, which complements the Main Theorem of Class Field Theory (1.2.2).

COMPLEMENTARY THEOREM (1.2.5). *Let L be a finite abelian extension of K, and consider the composite map*

$$C_K \xrightarrow{(1.2.2)} \operatorname{Gal}(K^{\mathrm{ab}}/K) \xrightarrow{\mathrm{proj.}} \operatorname{Gal}(L/K).$$

Let v be a finite place of K.

(i) *The place v splits completely in L/K if and only if the image of $K_v^\times$ by the composite map*

$$K_v^\times \to C_K \to \operatorname{Gal}(L/K)$$

is trivial (the first arrow is "inclusion into the v-component").

(ii) *The place v is unramified in L/K if and only if the image of V_v by the composite map*

$$K_v^\times \to C_K \to \operatorname{Gal}(L/K)$$

is trivial.

1.3. Explanation. We shall explain the facts in §1.1 by way of the abstract theorems of class field theory in §1.2. Let K be an algebraic number field and L a finite abelian extension. The composition $C_K \to \operatorname{Gal}(K^{\mathrm{ab}}/K) \to \operatorname{Gal}(L/K)$ factors through the following quotient of C_K:

$$(1.3.1) \qquad \frac{(\bigoplus_{v:\mathrm{unr.}} K_v^\times/V_v) \oplus (\bigoplus_{v:\mathrm{ram.}} K_v^\times) \oplus (\bigoplus_{v:\mathrm{real}} \mathbb{R}^\times/\mathbb{R}_+^\times)}{\operatorname{Image}(K^\times)}.$$

Here, v: unr. (resp. v: ram.) means that the direct sum is over all finite places of K that are unramified (resp. ramified) in L/K; v: real means that the direct sum is over all real places of K; we set $\mathbb{R}_+^\times = \{a \in \mathbb{R}^\times; \ a > 0\}$; $\operatorname{Image}(K^\times)$ is the image of $K^\times$ by the diagonal embedding. The reason why the map $C_K \to \mathrm{Gal}(L/K)$ factors through the quotient (1.3.1) is that, by (1.2.5)(ii), V_v vanishes in $\mathrm{Gal}(L/K)$ if v: unr., and $\mathbb{R}_+^\times$ for v: real (resp. the whole $K_v^\times \simeq \mathbb{C}^\times$ for v: complex) vanishes in $\mathrm{Gal}(L/K)$, being a connected group. Also, for any finite place v, there is a canonical isomorphism

$$K_v^\times / V_v \simeq \mathbb{Z},$$

with which the composite map $K_v^\times \to K_v^\times/V_v \simeq \mathbb{Z}$ assigns $a \in K_v^\times$ to the index of the prime ideal v appearing in the prime ideal decomposition of the ideal ao_K.

Let us consider the example $K = \mathbb{Q}$, $L = \mathbb{Q}(\sqrt{5})$ of (1.1.2). One verifies that the only prime ideal that is ramified in L/K is $5\mathbb{Z}$ (assume we know this). Now let p be a prime number different from 5. We shall employ the general theory of §1.2 to deduce the law:

$p\mathbb{Z}$ splits completely in $L/K \Leftrightarrow p \equiv 1$ or $4 \bmod 5$,

which was given in (1.1.2). Consider the diagonal image of $p \in \mathbb{Q}^\times$ in the numerator of (1.3.1):

$$\left(\bigoplus_{\ell:\text{prime},\neq 5} \mathbb{Z}\right) \oplus \mathbb{Q}_5^\times \oplus \mathbb{R}^\times/\mathbb{R}_+^\times.$$

Its ℓ-factor $\in \mathbb{Z}$ (for $\ell \neq 5$) is 1 if $\ell = p$ and is 0 if $\ell \neq p$; its $\mathbb{Q}_5^\times$-factor is p; and its $(\mathbb{R}^\times/\mathbb{R}_+^\times)$-factor vanishes. Since the diagonal image $\mathrm{Image}(\mathbb{Q}^\times)$ vanishes in $\mathrm{Gal}(L/K)$, the product of the image in $\mathrm{Gal}(L/K)$ of the element 1 in $\mathbb{Z}$ at $\ell = p$ and the image in $\mathrm{Gal}(L/K)$ of the element p of $\mathbb{Q}_5^\times$ is the unit element. Then (1.2.5)(i) applies to show that

$p\mathbb{Z}$ splits completely $\Leftrightarrow p$ is mapped to the unit element

by $\mathbb{Q}_5^\times \to \mathrm{Gal}(L/K)$.

On the other hand, the image of $V_5 = \mathbb{Z}_5^\times$ in $\mathrm{Gal}(L/K)$ does not vanish because $5\mathbb{Z}$ is ramified in L/K ((1.2.5)(ii)). After a simple argument, the reader will find that

$$\mathbb{F}_5^\times/(\mathbb{F}_5^\times)^2 \xleftarrow{\simeq} \mathbb{Z}_5^\times/(\mathbb{Z}_5^\times)^2 \xrightarrow{\simeq} \mathrm{Gal}(L/K),$$

where $\mathbb{F}_5 = \mathbb{Z}/5\mathbb{Z} = \mathbb{Z}_5/5\mathbb{Z}_5$. Thus we deduce

$p\mathbb{Z}$ splits completely in $L/K \Leftrightarrow p \bmod 5 \in (\mathbb{F}_5^\times)^2$

(i.e., $p \equiv 1$ or $4 \bmod 5$).

The above arguments hold not only for $\mathbb{Q}(\sqrt{5})$ but also for the following cases exactly in the same manner, and we are led to Gauß' reciprocity law for quadratic residues (it is a prototype of class field theory): let q be an odd prime number, and put

$$q^* = \begin{cases} q & \text{if } q \equiv 1 \bmod 4, \\ -q & \text{if } q \equiv 3 \bmod 4. \end{cases}$$

Put $K = \mathbb{Q}$ and $L = \mathbb{Q}(\sqrt{q^*})$. Then it can be shown that only $q\mathbb{Z}$ is ramified in L/K. (If we take $\mathbb{Q}(\sqrt{q})/\mathbb{Q}$ for $q \equiv 3 \bmod 4$, the prime ideal $2\mathbb{Z}$ is ramified there as well as

$q\mathbb{Z}$. This is the reason why we take $\mathbb{Q}(\sqrt{q^*})$ rather than $\mathbb{Q}(\sqrt{q})$.) Exactly the same reasoning as for $\mathbb{Q}(\sqrt{5})/\mathbb{Q}$ shows the criterion that, for a prime number $p \neq q$,

(1.3.2) $$p\mathbb{Z} \text{ splits in } L/K \Leftrightarrow p \bmod q \in (\mathbb{F}_q^\times)^2.$$

Incidentally, Gauß' reciprocity law for quadratic residues is as follows: let p be an odd prime number. For an integer a not divisible by p, we denote by $\left(\frac{a}{p}\right)$ the image of $a \bmod p$ by the map $\mathbb{F}_p^\times \to \mathbb{F}_p^\times/(\mathbb{F}_p^\times)^2 \simeq \{\pm 1\}$.

(1.3.3) (Reciprocity Law for Quadratic Residues) If p and q are two distinct odd prime numbers, then we have

$$\left(\frac{q}{p}\right) \cdot \left(\frac{p}{q}\right) = (-1)^{(p-1)/2 \cdot (q-1)/2}.$$

(1.3.3) can be rewritten as $\left(\frac{q^*}{p}\right) = \left(\frac{p}{q}\right)$. If p is an odd prime number and a is an integer not divisible by p, it can be shown rather easily that

$$p \text{ splits completely in } \mathbb{Q}(\sqrt{a})/\mathbb{Q} \Leftrightarrow \left(\frac{a}{p}\right) = 1.$$

So let us assume this is known to us. Then the condition on the left-hand side of (1.3.2) can be written as $\left(\frac{q^*}{p}\right) = 1$, whereas the right-hand side of (1.3.2) means $\left(\frac{p}{q}\right) = 1$; thus we see that (1.3.2) means (1.3.3).

1.4. Local class field theory. Let K be a nondiscrete locally compact field. Then K is isomorphic to either $\mathbb{R}$, $\mathbb{C}$, a finite extension of the p-adic number field $\mathbb{Q}_p$ for some prime number p (these are the local fields that arise as local fields of algebraic number fields), or the formal power series field in one variable over a finite field. Such a field K is a so-called "special field" in (1.2.1), and the Galois group $\mathrm{Gal}(K^{\mathrm{ab}}/K)$ is approximated by the multiplicative group $K^\times$.

MAIN THEOREM OF LOCAL CLASS FIELD THEORY (1.4.1). *If K is as above, there exists a canonical homomorphism*

$$K^\times \to \mathrm{Gal}(K^{\mathrm{ab}}/K)$$

that is approximately an isomorphism.

Here, being "approximately an isomorphism" means, for one thing, that it establishes, by pulling back, a one-to-one correspondence between the open subgroups of $\mathrm{Gal}(K^{\mathrm{ab}}/K)$ and the open subgroups of $K^\times$ of finite indices. Also, it is continuous and has dense image, and is injective if K is not isomorphic to $\mathbb{R}$ or $\mathbb{C}$.

The relation of local class field theory to class field theory for algebraic number fields is as follows:

(1.4.2) Let K be an algebraic number field and let v be a place of K. Then the following diagram is commutative:

$$\begin{array}{ccc} K_v^\times & \xrightarrow{(1.4.1)} & \mathrm{Gal}((K_v)^{\mathrm{ab}}/K_v) \\ \text{into the } v\text{-component} \downarrow & & \downarrow \text{canonical map} \\ C_K & \xrightarrow{(1.2.2)} & \mathrm{Gal}(K^{\mathrm{ab}}/K). \end{array}$$

If L/K is a finite abelian extension, we have

$$K_v^\times \xrightarrow{\mathrm{surj.}} \mathrm{Gal}(K_v L/K_v) \hookrightarrow \mathrm{Gal}(L/K),$$

and hence

$$K_v L = K_v \Leftrightarrow K_v^\times \to \mathrm{Gal}(L/K) \text{ is the zero-map.}$$

It can be shown rather easily that the condition of the left-hand side is equivalent to the condition that v splits completely in L/K. Thus (1.2.5)(i) follows. As we have seen here, the problem on what happens to each place in L/K, such as (1.2.5), should be understood as belonging to local class field theory and its relationship with class field theory for algebraic number fields.

§2. Generalization

2.1. Introduction. We stated in (1.2.1) that, for a special field k, there exists a good theory of abelian extensions, that is, a theory in which $\mathrm{Gal}(k^{\mathrm{ab}}/k)$ is approximated by another group. The main content of §2 is that such fields are not only algebraic number fields and their local fields, but also fields that are finitely generated over prime fields, together with those fields that play the role of their "local fields", higher-dimensional local fields.

DEFINITION 2.1.1. Let $n \geq 0$. An *n-dimensional global field* is either a finitely generated field over the prime field $\mathbb{F}_p$ for some prime number p of transcendence degree n, or a finitely generated field over $\mathbb{Q}$ of transcendence degree $n - 1$.

A zero-dimensional global field is a finite field, a one-dimensional global field is either an algebraic number field or an algebraic function field in one variable over a finite field, ..., and

$$\{\text{finitely generated fields over prime fields}\} = \bigcup_{n \geq 0} \{n\text{-dimensional global fields}\}.$$

DEFINITION 2.1.2. Let $n \geq 0$. A field K is said to be an *n-dimensional local field* if there is given a sequence $k_0, \ldots, k_n$ of fields that satisfies the following (i), (ii), (iii):

(i) k_0 is a finite field;
(ii) for $1 \leq i \leq n$, k_i is a complete discrete valuation field with residue field k_{i-1};
(iii) $k_n = K$.

A zero-dimensional local field is a finite field, and a one-dimensional local field is either a finite extension of $\mathbb{Q}_p$ for some prime number p or a formal power series field in one variable over a finite field.

What was stated in §1 and the simple theory of abelian extensions of finite fields, which we explain below, may be put uniformly as follows:

TABLE

field k	group that approximates $\mathrm{Gal}(k^{\mathrm{ab}}/k)$
0-dimensional local field	$\mathbb{Z} = K_0(k)$
1-dimensional local field	$k^\times = K_1(k)$
1-dimensional global field	$C_k = (\underline{\coprod}_v K_1(k_v))/K_1(k)$

Here, for a finite field $k = \mathbb{F}_q$ (q is a power of a prime number), when we say that $\mathrm{Gal}(k^{\mathrm{ab}}/k)$ is approximated by $\mathbb{Z}$ we mean that the group

$$\mathrm{Gal}(k^{\mathrm{ab}}/k) = \varprojlim_n \mathrm{Gal}(\mathbb{F}_{q^n}/\mathbb{F}_q) \simeq \widehat{\mathbb{Z}}$$

(where $\widehat{\mathbb{Z}} = \varprojlim_n \mathbb{Z}/n\mathbb{Z}$) is approximated by its dense subgroup $\mathbb{Z} \subset \widehat{\mathbb{Z}}$. According to algebraic K-theory, for any ring A, abelian groups $K_n(A)$ ($n = 0, 1, 2, \ldots$), called the K-groups of A, are defined, and for any field k, one has

$$K_0(k) = \mathbb{Z}, \qquad K_1(k) = k^\times.$$

Hence, the above table suggests that $\mathrm{Gal}(k^{\mathrm{ab}}/k)$ might be approximated by $K_2(k)$ if k is a two-dimensional local field; $\mathrm{Gal}(k^{\mathrm{ab}}/k)$ might be approximated by a sort of idele class group involving K_2 if k is a two-dimensional global field, etc.

Indeed, such is the truth. However, what are suitable for generalization of class field theory are not Quillen's K-groups, which are usually used as K-groups in algebraic K-theory (we mean Quillen's K-groups when we simply say K-groups and denote by K_n), but Milnor's K-groups, which are defined for fields and are a bit smaller than Quillen's K-groups. (Milnor's K-groups are denoted by K_n^M to be distinguished from Quillen's K-groups. They coincide with Quillen's K-groups if $n \leq 2$.)

2.2. Main theorems. For a field k, its nth Milnor K-group $K_n^M(k)$ is defined as follows ([45]):

$$K_0^M(k) = \mathbb{Z}, \qquad K_1^M(k) = k^\times,$$

and for $n \geq 2$, $K_n^M(k)$ is the abelian group defined by the following generators and relations (i), (ii) (group operation is denoted additively):

Generators: symbols $\{a_1, \ldots, a_n\}$ ($a_1, \ldots, a_n \in k^\times$)
Relations: (i)

$$\{a_1, \ldots, a_{i-1}, bc, a_{i+1}, \ldots, a_n\}$$
$$= \{a_1, \ldots, a_{i-1}, b, a_{i+1}, \ldots, a_n\} + \{a_1, \ldots, a_{i-1}, c, a_{i+1}, \ldots, a_n\}$$

$$(1 \leq i \leq n).$$

(ii) If $a_i + a_j = 1$ for some indices $i \neq j$, then $\{a_1, \ldots, a_n\} = 0$.
(It follows automatically from these relations (i) and (ii) that $\{a_1, \ldots, a_n\}$ is antisymmetric with respect to $a_1, \ldots, a_n$.)

The main theorem of the generalized local class field theory is as follows.

THEOREM 2.2.1. *Let K be an n-dimensional local field. Then there exists a canonical homomorphism*

$$K_n^M(K) \to \mathrm{Gal}(K^{\mathrm{ab}}/K)$$

that is approximately an isomorphism.

The meaning of being "approximately an isomorphism" is not very simple, because it seems impossible, if $n \geq 3$, to furnish the Milnor K-groups of K with a natural structure of topological groups. There is, however, a notion of "open subgroups" for Milnor K-groups (this "open" is not the "open" in the sense of general topology), and the open subgroups of $\mathrm{Gal}(K^{\mathrm{ab}}/K)$ and the open subgroups of $K_n^M(K)$ of finite index are in one-to-one correspondence by pulling back by the above canonical homomorphism ([9]).

Generalization of the class field theory for algebraic number fields is a joint work with Shuji Saito ([15]). Also, the paper [1] of Spencer Bloch (in which he studied class field theory of two-dimensional global fields) played a crucial role in our generalization.

Let K be an n-dimensional global field, and choose an integral scheme X with function field K that is projective over $\mathbb{Z}$ (then X is a scheme of dimension n). For example, if K is a number field, one can take $\mathrm{Spec}(o_K)$ as X. As we shall explain in §2.3, for a nonzero coherent ideal I of $\mathscr{O}_X$, we can define an abelian group $C_I(X)$ using nth Milnor K-groups, and if $I' \subset I$, there is defined a canonical surjection $C_{I'}(X) \to C_I(X)$. Set

$$C(X) = \varprojlim_I C_I(X),$$

where the inverse limit is taken as I shrinks. We give $C(X)$ the inverse limit topology, regarding each $C_I(X)$ as discrete. If K is an algebraic number field, we have

$$C(X) \simeq C_K/(\text{connected component of the unit element})$$

as topological groups.

The main theorem of generalization of class field theory for algebraic number fields is as follows.

THEOREM 2.2.2. *Let K be an n-dimensional global field, and choose an integral scheme X with function field K that is projective over $\mathbb{Z}$.*

(i) *If K is of characteristic 0, then each $C_I(X)$ is finite and we have*

$$C(X) \simeq \mathrm{Gal}(K^{\mathrm{ab}}/K) \quad (\text{as topological groups}).$$

(ii) *If K is of characteristic $p > 0$, then each $C_I(X)$ is of the form $\mathbb{Z} \oplus (\text{finite group})$, and we have*

$$C(X) \simeq \mathrm{Gal}(K^{\mathrm{ab}}/K) \times_{\widehat{\mathbb{Z}}} \mathbb{Z} \quad (\text{fiber product of topological groups}).$$

Here, $\widehat{\mathbb{Z}}$ is identified with $\mathrm{Gal}(\mathbb{F}_p^{\mathrm{ab}}/\mathbb{F}_p)$ and $\mathbb{Z}$ is regarded as discrete.

2.3. Idele class groups. We give in (i) – (v) below the definition and some properties of the group $C_I(X)$, which appeared in the above Theorem (2.2.2):

(i) *Cohomological definition*

First we define $C_I(X)$ as a cohomology group. A definition by explicit presentation, not abstract as this one, is also possible; it will make $C_I(X)$ look more like an idele class group (see (iv)). Let K be an n-dimensional global field, and X an integral scheme with function field K that is projective over $\mathbb{Z}$. If K does not have a structure of ordered field (for example, if the characteristic of K is not 0), the definition of $C_I(X)$ is simply

$$C_I(X) = H^n(X, K_n^M(\mathcal{O}_X, I)).$$

Here, $K_n^M(\mathcal{O}_X, I)$ is a sheaf of abelian groups in the Zariski topology, which is defined to be the subsheaf of the constant sheaf $K_n^M(K)$ generated by local sections of the form $\{a_1, \ldots, a_n\}$ ($a_1, \ldots, a_n \in \mathcal{O}_X^\times$, $a_1 \equiv 1 \bmod I$). If K has a structure of ordered field (such a structure corresponds to a real place in the case of algebraic number fields), we must modify $H^n(X, K_n^M(\mathcal{O}_X, I))$ somehow, accounting for what corresponds to the contribution of real places in the case of algebraic number fields (this will be explained later).

In the case where K is an algebraic number field and $X = \mathrm{Spec}(o_K)$, if we identify the sheaf I with an ideal of o_K, then $K_1^M(\mathcal{O}_X, I)$ (resp. $C_I(X)$) is what is called the ideal class group with conductor I (resp. with conductor equal to the product of all real places of K and I), which has been of significant importance in the class field theory of algebraic number fields (cf. [55]). The important point in the higher-dimensional case is that the counterparts of ideal class groups (without conductor) in the case of algebraic number fields are Chow groups $CH_0(X)$ of 0-cycles. Thus $H^n(X, K_n^M(\mathcal{O}_X, I))$ and $C_I(X)$ are, so to speak, CH_0 with conductor.

Now we give the definition of $C_I(X)$ in the case where K may have a structure of an ordered field, which will be somewhat complicated. Set $X_\mathbb{R} = X \otimes_\mathbb{Z} \mathbb{R}$, $I_\mathbb{R} = I \otimes_\mathbb{Z} \mathbb{R}$, and

$$S_I = H^{n-1}(X_\mathbb{R}, K_n^M(\mathcal{O}_{X_\mathbb{R}}, I_\mathbb{R})).$$

Define $C_I(X)$ to be the cokernel of the map

$$(2.3.1) \quad \bigoplus_{y \in X_1} H_y^{n-1}(X, K_n^M(\mathcal{O}_X, I)) \to \left(\bigoplus_{x \in X_0} H_x^n(X, K_n^M(\mathcal{O}_X, I)) \right) \oplus S_I/2S_I.$$

(Here, X_i denotes the set $\{x \in X ; \dim(\overline{\{x\}}) = i\}$ of points on X of dimension i, where $\overline{\{x\}}$ is the closure of $\{x\}$.) In general, for a sheaf of abelian groups $\mathscr{F}$ on X, we have

$$\bigoplus_{y \in X_1} H_y^{n-1}(X, \mathscr{F}) \to \bigoplus_{x \in X_0} H_x^n(X, \mathscr{F}) \to H^n(X, \mathscr{F}) \to 0 \quad \text{(exact)}.$$

Hence,

$$S_I/2S_I \to C_I(X) \to H^n(X, K_n^M(\mathcal{O}_X, I)) \to 0 \quad \text{(exact)}.$$

It can be proved that $S_I/2S_I$ is a finite group and that, if K does not have a structure of an ordered field, it is 0.

In this article, we work mainly with $C(X)$, which is the generalization of $C_K/$ (connected component of the unit element) in the case of algebraic number fields. If the reader should want to generalize C_K itself, he should take the $\varprojlim_I$ of the cokernel of (2.3.1) with $S_I/2S_I$ replaced by S_I (it coincides with C_K if K is an algebraic number field).

(ii) *Relation with CH_0*

For an integral scheme X, let $K_n^M(\mathscr{O}_X)$ be the subsheaf of the constant sheaf $K_n^M(K)$ (K is the function field of X) generated by all local sections $\{a_1, \ldots, a_n\}$ with $a_1, \ldots, a_n \in \mathscr{O}_X^\times$ (so $K_n^M(\mathscr{O}_X) = K_n^M(\mathscr{O}_X, \mathscr{O}_X)$). Then we have the following.

THEOREM 2.3.2. *If X is smooth over a field or an excellent Dedekind domain (for example, the integer ring of an algebraic number field) and $n = \dim(X)$, then*

$$H^n(X, K_n^M(\mathscr{O}_X)) \simeq CH^n(X).$$

(In general, for a Noetherian scheme S, we define its Chow groups by

$$CH^i(S) = \mathrm{Coker}\left(\bigoplus_{y \in S^{i-1}} \kappa(y)^\times \xrightarrow{\partial} \bigoplus_{x \in S^i} \mathbb{Z}\right),$$

$$CH_i(S) = \mathrm{Coker}\left(\bigoplus_{y \in S_{i+1}} \kappa(y)^\times \xrightarrow{\partial} \bigoplus_{x \in S_i} \mathbb{Z}\right),$$

where S^i is the set of points on S of codimension i, $\kappa(y)$ is the residue field of the point y, and ∂ is the homomorphism of which the (x, y)-component $\partial_{x,y}$ is defined as follows: if $x \notin \overline{\{y\}}$, then $\partial_{x,y}$ is the 0-map; if $x \in \overline{\{y\}}$, then let A be the local ring at x of $\overline{\{y\}}$ viewed as a reduced scheme and define

$$\partial_{x,y}(ab^{-1}) = \mathrm{length}_A(A/aA) - \mathrm{length}_A(A/bA)$$

for $a, b \in A - \{0\}$. If S is irreducible and of finite type over a field or $\mathbb{Z}$, then $CH^i(S) = CH_{\dim(S)-i}(S)$. If $S = \mathrm{Spec}(o_K)$ with K an algebraic number field, then

$$CH_0(S) = CH^1(S) \simeq \text{(ideal class group of } o_K).)$$

For such relations, as in (2.3.2), of Chow groups and algebraic K-theory, it is proved for Quillen's K-groups in Quillen [48] that

$$H^i(X, K_i(\mathscr{O}_X)) \simeq CH^i(X) \quad \text{for all } i \in \mathbb{Z}$$

for a smooth scheme X over a field (in the case $i = 2$, this is due to Bloch [25]). As for Milnor K-groups, see [36] and [54].

Using the finiteness of $C_I(X)$ in Theorem (2.2.2) together with (2.3.2), one can prove the next theorem.

THEOREM 2.3.3. *If X is a scheme of finite type over $\mathbb{Z}$, then $CH_0(X)$ is a finitely generated abelian group.*

In general, for a scheme of finite type over $\mathbb{Z}$, its Chow groups are conjectured to be finitely generated in all degrees. For CH^1, this is true, being reduced to the finite generation of rational points on abelian varieties over algebraic number fields (the Mordell-Weil theorem). The cases other than CH^0, CH^1, and CH_0 remain open.

(iii) *Relation with generalized Albanese varieties*

Let us consider what $H^n(X, K_n^M(\mathscr{O}_X, I))$ is in the case where X is a smooth proper variety of dimension n over a field k. The group $CH_0(X)^{\deg.0}$ (deg.0 means "the part of degree 0") is related to the Albanese variety Alb_X via the Albanese map $\alpha_X : CH_0(X)^{\deg.0} \to \mathrm{Alb}_X(k)$ ((k) denotes the group of k-rational points). One can say that $H^n(X, K_n^M(\mathscr{O}_X, I))$, which generalizes $CH_0(X)$, may be compared with a generalized Albanese variety which generalizes the Albanese variety (and is at the same time a higher-dimensional generalization of a generalized Jacobian variety of a curve; it is something like "Albanese variety with conductor" [24]). Indeed, one has the following commutative diagram:

$$\begin{array}{ccc} \varprojlim_{I} H^n(X, K_n^M(\mathcal{O}_X, I))^{\deg.0} & \xrightarrow{\tilde{\alpha}_X} & A(k) \\ {\scriptstyle\text{(surjective)}}\downarrow & & \downarrow \\ CH_0(X)^{\deg.0} & \xrightarrow{\alpha_X} & \mathrm{Alb}_X(k), \end{array}$$

where A is the inverse limit of the generalized Albanese varieties. If k is a finite field, using Theorem (2.2.2), one can prove that $\tilde{\alpha}_X$ is an isomorphism [14]. (The map $\tilde{\alpha}_X$ is not necessarily an isomorphism for a general field k.)

(iv) *Explicit presentation of idele class groups, relation of higher-dimensional local fields and higher-dimensional global fields*

Let K, X, and n be as in (i). By formal consideration on local cohomology, it can be seen that the map (2.3.1), which has cokernel $C_I(X)$, is, if $\dim(X) = 1$,

$$K^\times \to \left(\bigoplus_{x \in X_0} K^\times / K_1^M(\mathcal{O}_{X,x}, I_x) \right) \oplus S_I / 2S_I$$

(in this case, $S_I/2S_I = \bigoplus_{v:\text{real}} \mathbb{R}^\times / \mathbb{R}_+^\times$), and if $\dim(X) = 2$,

$$\bigoplus_{y \in X_1} K_2(K)/K_2^M(\mathcal{O}_{X,y}, I_y) \to \left(\bigoplus_{x \in X_0} \frac{\bigoplus_{y \in X_1, x \in \overline{\{y\}}} K_2(K)/K_2^M(\mathcal{O}_{X,y}, I_y)}{\mathrm{Image}(K_2(K))} \right) \oplus S_I/2S_I$$

(there are explicit presentations for $n \geq 3$ as well, which we omit since they become increasingly more complicated; cf. [15], §1).

We now employ the above explict presentation of $C_I(X)$ to explain the relation of class field theory for higher-dimensional local fields and class field theory for higher-dimensional global fields, taking up the case of dimension two as an example. From an n-dimensional global field K, one can obtain, as its "local fields", n-dimensional local fields as follows: pick an $x \in X_0$, pick a prime ideal $\mathfrak{p}$ of dimension one of the completion $\widehat{\mathcal{O}}_{X,x}$ of the local ring $\mathcal{O}_{X,x}$, pick a prime ideal $\mathfrak{p}'$ of dimension one of the completion $(\widehat{\mathcal{O}}_{X,x})_{\widehat{\mathfrak{p}}}$ of the local ring $(\widehat{\mathcal{O}}_{X,x})_{\mathfrak{p}}$, ..., and repeat this procedure as long as possible. Then one obtains an n-dimensional local field. Let $n = 2$ and, for simplicity, assume K does not have a structure of an ordered field and that X is normal. We replace the Zariski-local rings in the above presentation of $C_I(X)$ by their completions to define $\widehat{C}_I(X)$ as the cokernel of the map

$$\bigoplus_{y \in X_1} K_2(K_y)/K_2^M(\widehat{\mathcal{O}}_{X,y}, I\widehat{\mathcal{O}}_{X,y}) \to \bigoplus_{x \in X_0} \frac{\bigoplus_{\mathfrak{p}} K_2(K_{x,\mathfrak{p}})/K_2^M((\widehat{\mathcal{O}}_{X,y})_{\widehat{\mathfrak{p}}}, I(\widehat{\mathcal{O}}_{X,y})_{\widehat{\mathfrak{p}}})}{\mathrm{Image}(K_2(K_x))}.$$

(Here, K_y is the fraction field of $\widehat{\mathcal{O}}_{X,y}$; K_x is the fraction field of $\widehat{\mathcal{O}}_{X,x}$; for each $x \in X_0$, $\mathfrak{p}$ ranges over prime ideals of $\widehat{\mathcal{O}}_{X,x}$ of dimension 1; and $K_{x,\mathfrak{p}}$ is the fraction field of $(\widehat{\mathcal{O}}_{X,x})_{\widehat{\mathfrak{p}}}$, which is a two-dimensional local field.) Then the canonical map $C_I(X) \to \widehat{C}_I(X)$ is an isomorphism, and the commutative diagram

$$\begin{array}{ccc} K_2(K_{x,\mathfrak{p}}) & \longrightarrow & \text{Gal}((K_{x,\mathfrak{p}})^{\text{ab}}/K_{x,\mathfrak{p}}) \\ \downarrow & & \downarrow \\ C(X) \simeq \varprojlim_I \widehat{C}_I(X) & \longrightarrow & \text{Gal}(K^{\text{ab}}/K) \end{array}$$

embodies the relation between two-dimensional local class field theory and two-dimensional global class field theory (analogue of (1.4.2)).

It is also possible to follow the style in the case of algebraic number fields to define the idele class group of K in the form

$$\coprod_{(x,\mathfrak{p})} K_2(K_{x,\mathfrak{p}}) / \text{Image} \left(\coprod_{x \in X_0} K_2(K_x) \times \coprod_{y \in X_1} K_2(K_y) \right)$$

(with an appropriate definition of $\coprod$) and understand $C_I(X) = \widehat{C}_I(X)$ as a quotient of it.

(v) *Relation with Frobenius substitutions*

Let K, X, and n be as in (i). Suppose U is a dense open subset of X and is smooth over a finite field or the integer ring of an algebraic number field. Then we have the following:

(2.3.4) If $I|_U = \mathcal{O}_U$, there exists a canonical isomorphism $H^n_x(X, K^M_n(\mathcal{O}_X, I)) \simeq \mathbb{Z}$ for each $x \in U_0$, and the map

$$\bigoplus_{x \in U_0} \mathbb{Z} \to C_I(X)$$

is surjective.

(2.3.5) If L/K is a finite abelian extension that is unramified over U, then the composite map $C(X) \to \text{Gal}(K^{\text{ab}}/K) \to \text{Gal}(L/K)$ factors through $C_I(X)$ for some coherent ideal I such that $I|_U = \mathcal{O}_U$, and the image in $C_I(X)$ by the map of (2.3.4) of $1 \in \mathbb{Z}$ at $x \in U_0$ is mapped by the map $C_I(X) \to \text{Gal}(L/K)$ to the Frobenius substitution of x.

The decomposition of a point $x \in U_0$ in L/K is determined by the Frobenius substitution. Thus it follows from (2.3.5) that it is determined by the class of x in $C_I(X)$ (i.e., the image in $C_I(X)$ of $1 \in \mathbb{Z}$ at x). (This fact corresponds to the existence of laws as were given in (1.1) for algebraic number fields.)

S. Lang ([17]) was the first to note that the classical definition of Frobenius substitutions generalizes to the higher-dimensional case and tried to extend class field theory to the case of function fields in several variables over a finite field. A. N. Paršin ([19], [20]) started K-theoretic generalization of class field theory independently of the author (in fact, a little earlier than the author).

§3. Supplements

We shall discuss various themes to which we are necessarily led in pursuing a generalization of class field theory.

3.1. Milnor K-groups of complete discrete valuation fields.

In §2.2, we said that $\mathrm{Gal}(K^{\mathrm{ab}}/K)$ is approximately isomorphic to $K_n^M(K)$ if K is an n-dimensional local field. Hence we want to understand $K_n^M(K)$ of such fields.

In this subsection, let K be a general complete discrete valuation field and F its residue field. We shall consider how Milnor K-groups of K are described in terms of F. The important point is that the differential forms of the residue field come into the game.

For $K_1^M(K) = K^\times$, it is well known that it has a decreasing filtration by the ith unit group $U_K^{(i)} \subset K^\times$ ($i \geq 0$) (we define $U_K^{(i)} = \mathrm{Ker}((o_K)^\times \to (o_K/m_K^i)^\times)$, where o_K is the valuation ring of K and m_K is its maximal ideal), and that

(i) $K^\times/U_K^{(0)} \simeq \mathbb{Z}$, (ii) $U_K^{(0)}/U_K^{(1)} \simeq F^\times$, (iii) $U_K^{(i)}/U_K^{(i+1)} \simeq F$ for $i \geq 1$.

Similarly, $K_n^M(K)$, $n \geq 1$, has a good filtration $U^i K_n^M(K)$ ($i \geq 0$). Define the unit part $U^0 K_n^M(K)$ of $K_n^M(K)$ to be the subgroup generated by $\{a_1, \ldots, a_n\}$ ($a_1, \ldots, a_n \in U_K^{(0)}$), and define $U^i K_n^M(K)$ for $i \geq 1$ to be the subgroup generated by $\{a_1, \ldots, a_n\}$ ($a_1, \ldots, a_n \in K^\times$, $a_1 \in U_K^{(i)}$). (Then $U^0 K_n^M(K) \supset U^1 K_n^M(K)$.) First, we have ([45])

(i) $K_n^M(K)/U^0 K_n^M(K) \simeq K_{n-1}^M(F)$,

(ii) $U^0 K_n^M(K)/U^1 K_n^M(K) \simeq K_n^M(F)$.

The problem is therefore to know $\mathrm{gr}^i K_n^M(K) = U^i K_n^M(K)/U^{i+1} K_n^M(K)$, $i \geq 1$. (N.B. $K_n^M(K)$ intervenes in number theory always through $\varprojlim_i K_n^M(K)/U^i K_n^M(K)$; we ignore $\bigcap_{i \geq 0} U^i K_n^M(K)$ for the moment.) If K is an n-dimensional local field with $n \geq 1$, then the isomorphism of (i) corresponds to the isomorphism $\mathrm{Gal}(K^{\mathrm{abnr}}/K) \simeq \mathrm{Gal}(F^{\mathrm{ab}}/F)$ (where K^{abnr} is the maximal unramified abelian extension of K) and, similarly as in the case of one-dimensional local class field theory, $U^0 K_n^M(K)$ approximates the ramification group of $\mathrm{Gal}(K^{\mathrm{ab}}/K)$ and $U^1 K_n^M(K)$ approximates its wild part.

In the following, we denote by $\Omega_{F/\mathbb{Z}}^q = \wedge_F^q \Omega_{F/\mathbb{Z}}^1$ the module of differential forms of degree q of the field F.

(iii) Fix a prime element π of K. For each $i \geq 1$, there exists the following surjective homomorphism:

$$\varphi_{\pi,i} : \Omega_{F/\mathbb{Z}}^{n-1} \oplus \Omega_{F/\mathbb{Z}}^{n-2} \to \mathrm{gr}^i K_n^M(K),$$

$$(a\frac{db_1}{b_1} \wedge \cdots \wedge \frac{db_{n-1}}{b_{n-1}}, 0) \mapsto \{1 + \pi^i \tilde{a}, \tilde{b}_1, \ldots, \tilde{b}_{n-1}\},$$

$$(0, a\frac{db_1}{b_1} \wedge \cdots \wedge \frac{db_{n-2}}{b_{n-2}}) \mapsto \{1 + \pi^i \tilde{a}, \tilde{b}_1, \ldots, \tilde{b}_{n-2}, \pi\}.$$

(Here $\tilde{a}$ and $\tilde{b}_i$ are arbitrary liftings to o_K of a and b_i, respectively.)

Thus gr^i, for $i \geq 1$, will be understood by knowing the kernel of the surjection $\varphi_{\pi,i}$. This kernel is known if K is of equal characteristic (i.e., $\mathrm{char}(K) = \mathrm{char}(F)$) ([26], [33]), but in the mixed characteristic case (i.e., $\mathrm{char}(K) = 0$ and $\mathrm{char}(F) = p > 0$), it is known, in general, only when $i \leq e_K p/(p-1)$ (e_K is the absolute ramification index $\mathrm{ord}_K(p)$) ([28], [41]). Since $U^i K_n^M(K) \subset p K_n^M(K)$ if $i > e_K p/(p-1)$, it follows that $\mathrm{gr}^i (K_n^M(K)/p K_n^M(K))$ is well understood for all i, where the gr is with respect to the quotient filtration on $K_n^M(K)/p K_n^M(K)$ induced from that on $K_n^M(K)$. Noticing what will be mentioned in §3.2 below, it seems to be a fundamental and important problem to have a better understanding of $K_n^M(K)$.

3.2. From the point of view of algebraic geometry over p-adic fields. The content of §3.1, in the case of mixed characteristic, has the following meaning in algebraic geometry over p-adic fields. First, it gives rise to the following *yoga* in our thought:

(3.2.1) Groups associated with p-adic objects in characteristic 0 are composed of a pile of differential modules in characteristic p.

(This idea was not recognized very explicitly in number theory of algebraic number fields, since the residue field F of a local field of an algebraic number field is perfect and hence $\Omega^1_{F/\mathbb{Z}} = 0$.)

Let V be a variety over $\mathbb{Q}_p$, X a normal model of V over $\mathbb{Z}_p$, and Y the reduction mod p of X. If K is the fraction field of the completion $\widehat{\mathcal{O}}_{X,\mathfrak{p}}$ of the local ring $\mathcal{O}_{X,\mathfrak{p}}$ of X at a generic point $\mathfrak{p}$ of Y, then K is a complete discrete valuation field and its residue field F is the function field of an irreducible component of Y. The differential module $\Omega^q_{F/\mathbb{Z}}$ is exactly the rational differential q-form on this irreducible component. Thus, studying a complete discrete valuation field whose residue field may not be perfect, as in §3.1, means studying algebraic geometry over a p-adic field generically along the reduction Y. Then, geometrically, what does the study, as in §3.1, of Milnor K-groups mean? They are related to the important objects, p-adic étale cohomology groups of the variety V over a p-adic field, via the following theorem.

THEOREM 3.2.2 ([28, §5]). *Let K be a complete discrete valuation field of characteristic 0 whose residue field is of characteristic $p > 0$. Then for all q and n, we have*
$$K_q^M(K)/p^n K_q^M(K) \simeq H^q(K, \mathbb{Z}/p^n\mathbb{Z}(q)).$$

Here $H^q(K, \ )$ is the qth Galois cohomology group of K where (q) denotes the qth Tate twist. The reason that it is related to p-adic étale cohomology is that the right-hand side of (3.2.2) is the p-adic étale cohomology of V considered locally at a generic point of Y. In particular, the facts in §3.1 about Milnor K-groups yield local results at generic points of Y on the relation of p-adic étale cohomology of V and differential forms on Y, which is an important subject of algebraic geometry over a p-adic field. For all this and its generalization, see [28] and [35].

REMARK. It is conjectured that, for a general field k and an integer n that is invertible in k, one should have
$$K_q^M(k)/n K_q^M(k) \simeq H^q(k, \mathbb{Z}/n\mathbb{Z}(q)).$$
If $q = 2$, this is a theorem due to Mercuriev-Suslin.

3.3. Galois cohomology and duality. To know the abelian extensions of a field k is to know $\mathrm{Gal}(k^{\mathrm{ab}}/k)$ (§2.1), but it is the same as knowing the Galois cohomology group $H^1(k, \mathbb{Q}/\mathbb{Z})$, which is the Pontrjagin dual of $\mathrm{Gal}(k^{\mathrm{ab}}/k)$. Further, it is natural to deal not only with H^1 but also with Galois cohomology groups of various degrees. In particular, the Galois cohomology groups $H^q(k, (\mathbb{Q}/\mathbb{Z})(q-1))$ ($q \in \mathbb{Z}$) are considered to be an important family. It is the dual group of $\mathrm{Gal}(k^{\mathrm{ab}}/k)$ if $q = 1$ and is the Brauer group $\mathrm{Br}(k)$ of k (= the set, with a group structure, of isomorphism classes of skew fields of finite degrees over k with center k). It is an important fact that $\mathrm{Br}(k) \simeq \mathbb{Q}/\mathbb{Z}$ if k is a one-dimensional local field; this means that $\mathrm{Br}(k)$ is approximately the dual group of $K_0^M(k) = \mathbb{Z}$. The generalization of this fact to higher-dimensional local fields is Theorem (3.3.1) below ([7], [8]). The case $q = 1$ in (3.3.1) is nothing but class field theory for higher-dimensional local fields. (For, to say that $\mathrm{Gal}(k^{\mathrm{ab}}/k)$ is approximately $K_n^M(k)$ is to say that $H^1(k, \mathbb{Q}/\mathbb{Z})$ is approximately the dual group of $K_n^M(k)$.)

In what follows, we write $H^q(k)$ for $H^q(k, (\mathbb{Q}/\mathbb{Z})(q-1))$. (In fact, the Tate twist does not work well in characteristic $p > 0$ for p-power torsion groups; so we have to give a good definition of the p-power torsion part of $H^q(k)$ for k of characteristic p in a separate way. But we do not go into the details now. See [7], II, §3.)

THEOREM 3.3.1. *Let K be an n-dimensional local field. For each $q \geq 0$, $H^q(K)$ is isomorphic to the group of continuous homomorphisms of $K_{n+1-q}^M(K)$ to $\mathbb{Q}/\mathbb{Z}$ of finite orders. In particular, we have $H^{n+1}(K) \simeq \mathbb{Q}/\mathbb{Z}$.*

Here, a homomorphism of a Milnor K-group of K to $\mathbb{Q}/\mathbb{Z}$ is said to be continuous if its kernel is open in the sense mentioned following (2.2.1).

Class field theory for an n-dimensional global field K can also be understood as a duality theorem

$$H^1(K) \simeq \mathrm{Hom}_{\mathrm{cont}}(C(X), \mathbb{Q}/\mathbb{Z}),$$

where $C(X)$ is as in §2.2. For an arbitrary $q \geq 0$, one can define the K_{n+1-q}^M-idele class group of K and a canonical homomorphism of $H^q(K)$ to the group of continuous homomorphisms of this idele class group to $\mathbb{Q}/\mathbb{Z}$. It is not known, however, for a general q, how perfect this duality is. As an example, for an algebraic number field K and its Brauer group $\mathrm{Br}(K)$, one has the classical and important theorem that

$$0 \to \mathrm{Br}(K) \to \left(\bigoplus_{v:\mathrm{finite}} \mathbb{Q}/\mathbb{Z}\right) \oplus \left(\bigoplus_{v:\mathrm{real}} \mathbb{Z}/2\mathbb{Z}\right) \to \mathbb{Q}/\mathbb{Z} \to 0 \quad \text{(exact)}.$$

This means that $\mathrm{Br}(K)$ is isomorphic to the group of continuous homomorphisms of the K_0^M-idele class group $(\prod_{v:\mathrm{finite}} \mathbb{Z} \times \prod_{v:\mathrm{real}} \mathbb{Z}/2\mathbb{Z})/\mathbb{Z}$ to $\mathbb{Q}/\mathbb{Z}$. If K is a two-dimensional global field, $H^3(K)$ is isomorphic to the group of continuous homomorphisms of the K_0^M-idele class group of K to $\mathbb{Q}/\mathbb{Z}$ ([10]).

Such dualities have already been studied by Manin ([44]) in the case $q = 2$, i.e., for $\mathrm{Br}(K)$. His formulation is a little different from the above; the duality has been understood, for a variety V over an algebraic number field F, as a duality between $\mathrm{Br}(V)$ and the "idele class group"

$$(\coprod_v CH_0(V \otimes_F F_v))/\mathrm{Image}(CH_0(V))$$

(v ranges over the places of F). In this direction, much has been studied by Colliot-Thélène and Sansuc (e.g. [30]).

We conclude this section with some remarks on H^q of a complete discrete valuation field. In the rest of this section, let K be a complete discrete valuation field and F its residue field.

$\prod_{\mathfrak{p}} \mathfrak{p}^{n(\mathfrak{p})}$ where $\mathfrak{p}$ ranges over the nonzero prime ideals of o_L, then we mean by the wild part of the different the element of $CH_0(F)$ given by

$$(n(\mathfrak{p}) - e_\mathfrak{p} + 1)_\mathfrak{p} \in \bigoplus_{\mathfrak{p} \in F} \mathbb{Z}.$$

($e_\mathfrak{p}$ is the ramification index of $\mathfrak{p}$.)

(ii) If V is a Galois covering with Galois group G, then a $CH_0(F)$-valued function $C(\sigma)$ on G ($\sigma \in G$), generalizing the Swan character, should be defined.

In the classical case, if $\sigma \neq 1$, the $\mathfrak{p}$-component of $C(\sigma) \in \bigoplus_{\mathfrak{p} \in F} \mathbb{Z}$ is

$$-\sup(\mathrm{length}_{\mathscr{O}_{Y,\mathfrak{p}}}(\mathscr{O}_{Y,\mathfrak{p}}/I_{\sigma,\mathfrak{p}}) - 1,\ 0),$$

where I_σ is the ideal of o_L generated by $\{\sigma(a) - a;\ a \in o_L\}$, and for $\sigma = 1$, we put $C(1) = -\sum_{\sigma \neq 1} C(\sigma)$.

(iii) Keep the assumption in (ii) and let ρ be a finite-dimensional $\mathbb{C}$-linear representation of G. Then an element $C(\rho) \in CH_0(E)$ generalizing the Swan conductor should be defined. $C(\rho)$ should depend only on the triple $(X, U, \rho : \pi_1(U) \to \mathrm{GL}_n(\mathbb{C}))$ and not on the coverings V and Y that trivialize ρ. If $\deg(\rho) = 1$, then $C(\rho)$ should be related to the generalized class field theory.

In the classical case, its relation with class field theory is as follows: the $\mathfrak{p}$-component of $C(\rho) \in \bigoplus_{\mathfrak{p} \in E} \mathbb{Z}$ is equal to the minimal integer $n \geq 0$ such that $U_\mathfrak{p}^{(n+1)}$ is annihilated by the map $K_\mathfrak{p}^\times \to \mathrm{Gal}(K^{\mathrm{ab}}/K) \xrightarrow{\rho} \mathbb{C}^\times$ induced by ρ (the first arrow is the reciprocity map of local class field theory).

Various formulas that hold for the objects in (i), (ii), (iii) in the classical case should generalize to formulas as equalities between elements of $CH_0(E)$ or $CH_0(F)$. If X is of finite type over $\mathbb{Z}$, then $C(\rho)$ should play the same role in the functional equations of L-functions $L(s, \rho)$ ([53]) as it played in the classical case.

4.2. Results. As for (i) and (ii) of the above (i), (ii), (iii), they are essentially done in Bloch [27] (precise results are given in the two-dimensional case). Though [27] is an important paper, we have no space here to describe its contents. In the following, we discuss the next theorem about (iii) above.

THEOREM 4.2.1. *If $\deg(\rho) = 1$ and $\dim(X) \leq 2$, then $C(\rho) \in CH_0(E)$ can be constructed.*

We shall construct $C(\rho)$ in §4.3, following the analogy with the theory of $\mathscr{D}$-modules. As we see below, this $C(\rho)$ is related to the solution of the two-dimensional case of a conjecture of Serre, to a Riemann-Roch formula for ℓ-adic sheaves, and to the conductor of the L-function of a degree-one representation of the Galois group of a two-dimensional global field.

(1) *A conjecture of Serre.* This is a very old conjecture on ramification in the higher-dimensional case ([51]). Let A be a regular local ring, and let G be a finite group consisting of automorphisms of A. Assume the following (a), (b), (c).

(a) For any $\sigma \in G - \{1\}$, let I_σ be the ideal of A generated by $\{\sigma(a) - a; a \in A\}$. Then $\mathrm{length}_A(A/I_\sigma)$ is finite.

(b) The G-fixed part A^G of A is a Noetherian ring.

(c) The local rings A^G and A have the same residue field.

Define a function $a_G : G \to \mathbb{Z}$ by $a_G(\sigma) = -\text{length}_A(A/I_\sigma)$ if $\sigma \in G - \{1\}$ and $a_G(1) = -\sum_{\sigma \neq 1} a_G(\sigma)$.

SERRE'S CONJECTURE. *a_G is the character of some representation of G.*

This conjecture is valid if $\dim(A) = 1$ (the representation is called the *Artin representation* and a_G is called the *Artin character*).

THEOREM 4.2.2. *Serre's conjecture holds true if $\dim(A) = 2$.*

To explain the relation of the conjecture to $C(\rho)$ in (4.2.1), quantities of Swan type are more convenient to deal with than quantities of Artin type. So we define a function $s_G : G \to \mathbb{Z}$ by $s_G(\sigma) = a_G(\sigma) + 1$ (for $\sigma \in G - \{1\}$) and $s_G(1) = -\sum_{\sigma \neq 1} s_G(\sigma)$, and consider Serre's conjecture in an equivalent form:

"s_G is the character of some representation of G".

Since a_G and s_G are known to be the characters of a representation once they are known to be the characters of a virtual representation, Serre's conjecture is equivalent to saying that, for any representation ρ of G of finite degree, the inner product of the character χ_ρ of ρ and s_G belongs to $\mathbb{Z}$;

$$(4.2.3) \qquad \frac{1}{[G]} \sum_{\sigma \in G} s_G(\sigma) \chi_\rho(\sigma) \in \mathbb{Z}.$$

($[G]$ is the order of G.)

Recall the proof of (4.2.3) in the case $\dim(A) = 1$. First it was reduced to the case $\deg(\rho) = 1$, in which case (4.2.3) is called the Hasse-Arf theorem and has a class-field-theoretical proof ([52]). As we declared at the beginning of this article, our belief is that traditional techniques and viewpoints of classical class field theory, in appropriate formulation, should apply to higher-dimensional cases. Indeed, we can reduce the proof of (4.2.2) for $\dim(A) = 2$ to the case $\deg(\rho) = 1$ imitating the methods in the classical case $\dim(A) = 1$. In the case $\deg(\rho) = 1$, assume that the residue field of A is perfect (we may do so), and consider the situation of §4.1 with $Y = \text{Spec}(A)$, $X = \text{Spec}(A^G)$, $V = \text{Spec}(A) - \{\text{closed point}\}$, $U = \text{Spec}(A^G) - \{\text{closed point}\}$, F and E are respectively the closed points of $\text{Spec}(A)$ and $\text{Spec}(A^G)$. We have

$$CH_0(E) \simeq \mathbb{Z}, \qquad CH_0(F) \simeq \mathbb{Z}.$$

If we identify these groups by the isomorphisms, we have $s_G(\sigma) = C(\sigma) \in \mathbb{Z}$. We can prove that, in the two-dimensional case, the element $C(\rho) \in \mathbb{Z}$, which we claimed in (4.2.1) to have constructed, has the same relation to $C(\sigma)$'s ($\sigma \in G$) as it had in the classical case and thus it coincides with the left-hand side of (4.2.3) (hence the left-hand side turns out to be an integer).

In this equality "the left-hand side of (4.2.3) = $C(\rho)$", as well as in the formulas (4.2.4) and (4.2.5) below, the definitions of the left and right sides are totally different, and this is the point that makes the formulas meaningful.

(2) *Riemann-Roch formula for ℓ-adic sheaves.* Let k be an algebraically closed field, X a smooth proper surface over k, and U a dense open subset of X. Let ℓ be a prime number different from $\text{char}(k)$, $\Lambda = \overline{\mathbb{Q}}_\ell$ or $\overline{\mathbb{F}}_\ell$ (the $^-$ means a fixed algebraic closure), and $\mathscr{F}$ a rank-one smooth sheaf of Λ-modules on U with respect to the étale topology ([31]). Then $\mathscr{F}$ corresponds to a representation $\rho : \pi_1(U) \to \Lambda^\times$, and as in the case of $\Lambda = \mathbb{C}$, one can define an element $C(\rho)$. We write $C(\mathscr{F})$ for it.

THEOREM 4.2.4. *We have*
$$\chi(U,\mathscr{F}) = \chi(U,\Lambda) - \deg(C(\mathscr{F})),$$
where we define $\chi(U,\mathscr{F}) = \sum_{q\in\mathbb{Z}}(-1)^q \dim_\Lambda H^q_{\text{et}}(U,\mathscr{F})$.

(See T. Saito [50b] for a generalization of (4.2.4) to higher-dimensional cases under certain assumptions.)

In addition to this theorem, there are other types of formulations ([43], [49]) of a higher-dimensional generalization of the Riemann-Roch formula for ℓ-adic sheaves on curves (= the Grothendieck-Ogg-Shafarevich formula); Grothendieck and Deligne also have other types of approach — the best method does not seem to have been established.

(3) *Conductor of L-functions.* Let A be a discrete valuation ring with perfect residue field k, N its fraction field, X a regular scheme that is proper, flat, of relative dimension one over A, and U a dense open subset of $X \otimes_A N$. Let ℓ be a prime number different from char(k), $\Lambda = \overline{\mathbb{Q}}_\ell$ or $\overline{\mathbb{F}}_\ell$, and let $\mathscr{F}$ be a rank-one smooth sheaf of Λ-modules on U with respect to the étale topology. We assume that the ramification of $\mathscr{F}$ over $X \otimes_A N$ is at worst tame (this assumption is automatically satisfied if char(N) = 0). For each $q \in \mathbb{Z}$, $H^q_{\text{et}}(U \otimes_N N^{\text{sep}}, \mathscr{F})$ is a Gal(N^{sep}/N)-module (N^{sep} is a separable closure of N). We write $\text{Sw}_A H^q_{\text{et}}(U \otimes_N N^{\text{sep}}, \mathscr{F})$ for the Swan conductor of this representation of Gal(N^{sep}/N) with respect to the discrete valuation ring A, and set

$$\text{Sw}_A R\Gamma(U \otimes_N N^{\text{sep}}, \mathscr{F}) = \sum_{q\in\mathbb{Z}}(-1)^q \text{Sw}_A H^q_{\text{et}}(U \otimes_N N^{\text{sep}}, \mathscr{F}).$$

Then the following formula holds.

THEOREM 4.2.5. $\text{Sw}_A R\Gamma(U \otimes_N N^{\text{sep}}, \mathscr{F}) = \text{Sw}_A R\Gamma(U \otimes_N N^{\text{sep}}, \Lambda) + \deg(C(\mathscr{F}))$.

On the other hand, a formula that expresses $\text{Sw}_A R\Gamma(U \otimes_N N^{\text{sep}}, \Lambda)$ in terms of differential forms is given in Bloch [27].

The meaning of Theorem (4.2.5) is the following: the ramification occurring on the two-dimensional scheme X (the ramification of $\mathscr{F}$) induces, via the operation of taking étale cohomology, ramification on the base scheme Spec(A), and the formula expresses the wildness Sw_A of the ramification induced on Spec(A) in terms of the quantity $C(\mathscr{F})$ of the initial ramification on X.

A more number-theoretic meaning is that (4.2.5) gives the "conductor" of the L-function $L(s,\rho)$ ([53]) of a representation Gal(K^{ab}/K) → $\mathbb{C}^\times$ of degree one for an algebraic function field K in one variable over an algebraic number field N. Regard ρ as a representation Gal(K^{ab}/K) → $\overline{\mathbb{Q}}_\ell^\times$ in an appropriate manner, and suppose it corresponds to a smooth sheaf $\mathscr{F}$ of $\overline{\mathbb{Q}}_\ell$-modules of rank one on an algebraic curve U over N. Then $L(s,\rho)$ coincides with the L-function of the virtual ℓ-adic representation $R\Gamma(U \otimes_N N^{\text{sep}}, \mathscr{F})$ of Gal(N^{sep}/N). If $\mathfrak{p}$ is a nonzero prime ideal of o_N and A is the local ring of o_N at $\mathfrak{p}$, then the left-hand side of (4.2.5) is the $\mathfrak{p}$-component of the conductor of $L(s,\rho)$ (i.e., the conductor part that should appear in the conjectural functional equation for $L(s,\rho)$), and thus the Theorem says that it is given by the right-hand side of (4.2.5).

EXAMPLE 4.2.6. Let N be an algebraic number field and set $K = N(T)$. Assume that N contains a primitive nth root of unity. We regard the function field L of the Fermat curve $T_0^n = T_1^n + T_2^n$ as an abelian extension of K with Galois group isomorphic to $\mathbb{Z}/n\mathbb{Z} \times \mathbb{Z}/n\mathbb{Z}$ by writing $L = K(\alpha, \beta)$ with $\alpha^n = T$ and $\beta^n = 1 - T$, where α is thought of as the function T_1/T_0 and β as T_2/T_0. Take $U = \mathbb{P}^1 - \{0, 1, \infty\}$. Then for each representation $\rho : \mathrm{Gal}(L/K) \to \mathbb{C}^\times$ of degree one, $R\Gamma(U \otimes_N N^{\mathrm{sep}}, \mathscr{F})$ is a virtual representation of $\mathrm{Gal}(N^{\mathrm{sep}}/N)$ of "Jacobi-sum type". The conductor of this representation (or of $L(s, \rho)$) is given by (4.2.5). The conductor of a Galois representation of Jacobi-sum type has already been calculated in [29]; we obtain another proof of this result in [29] by actually calculating $C(\rho)$ in the right-hand side of (4.2.5) with $X = \mathbb{P}^1_A$.

4.3. Analogy with $\mathscr{D}$-modules. In this subsection 4.3, we define the $C(\rho)$ mentioned in Theorem (4.2.1), following the analogy with the theory of $\mathscr{D}$-modules. In the theory of $\mathscr{D}$-modules, an important 0-cycle class is defined for a $\mathscr{D}$-module as the intersection of its characteristic cycle, which is defined on the cotangent bundle, and the 0-section of the cotangent bundle; it is this construction that we mimic.

Let X be a regular connected scheme, D a divisor with normal crossings on X, and $U = X - D$. In the rest of the paper, we proceed with our consideration comparing the following two cases.

In Case I, we consider r-dimensional continuous $\mathbb{C}$-linear representations of $\pi_1(U)$ over $\mathbb{C}$. We assume that the following condition is satisfied: the wild ramification locus E of ρ on X $(= \cup_{\mathfrak{p}} \overline{\{\mathfrak{p}\}}$, where $\mathfrak{p}$ ranges over the generic points of D at which ρ is wildly ramified; E is regarded as a reduced scheme) is a disjoint union of schemes of finite type over perfect fields.

In Case II, we let X be of finite type over a field k of characteristic 0, $\mathscr{D}_{U/k}$ the sheaf of rings of differential operators $\mathscr{O}_U \to \mathscr{O}_U$, and consider sheaves $\mathscr{M}$ of $\mathscr{D}_{U/k}$-modules on U that are locally free of finite rank r as $\mathscr{O}_U$-modules.

The following analogy between Case I and Case II is known:

ramification of ρ (on X, outside U) $\Leftrightarrow$ singularity of $\mathscr{M}$ (on X, outside U),

tame ramification of ρ $\Leftrightarrow$ regular singularity of $\mathscr{M}$,

wild ramification of ρ $\Leftrightarrow$ irregular singularity of $\mathscr{M}$.

We shall consider the case $r = 1$ in the following. In this case, the theory of $\mathscr{D}$-modules is not interesting, objects there being too simple, whereas Case I is interesting enough, and things in the general case are often reduced to the case $r = 1$. We shall construct analogues (i) – (v) in Case I below ($C(\rho)$ will be obtained in (v)), following the counterparts (i) – (v) in Case II.

First we consider Case II. Let $\Omega^1_{X/k}(\log D)$ be the sheaf of differential forms on X admitting logarithmic poles at D. Choose locally a basis e of the invertible $j_*(\mathscr{O}_U)$-module $j_*(\mathscr{M})$ ($j : U \hookrightarrow X$ is the inclusion map), and define a local section ω of $j_*\Omega^1_{U/k}$ by the equality $\nabla(e) = e \otimes \omega$ (∇ is the connection associated with the $\mathscr{D}$-module structure of $j_*(\mathscr{M})$). This ω depends on the choice of e, but the following objects are well defined.

(i) Define a divisor $\underline{\mathrm{Irr}}(\mathscr{M})$ on X by

$$\underline{\mathrm{Irr}}(\mathscr{M}) = \sum_{\mathfrak{p}} \mathrm{Irr}_{\mathfrak{p}}(\mathscr{M}) \cdot \mathfrak{p},$$

where the sum is over the generic points $\mathfrak{p}$ of D, and

$$\mathrm{Irr}_{\mathfrak{p}}(\mathscr{M}) = \min\{n \geq 0; \, \omega_{\mathfrak{p}} \in m_{X,\mathfrak{p}}^{-n} \Omega^1_{X/k}(\log D)_{\mathfrak{p}}\}.$$

Here $m_{X,\mathfrak{p}}$ is the maximal ideal of $\mathcal{O}_{X,\mathfrak{p}}$.

Thus ω is a local section of $\Omega^1_{X/k}(\log D)(\underline{\mathrm{Irr}}(\mathcal{M}))$. (Here the notation $\underline{\mathrm{Irr}}(\mathcal{M})$ means tensoring over $\mathcal{O}_X$ the invertible sheaf $\mathcal{O}_X(\underline{\mathrm{Irr}}(\mathcal{M}))$ associated with the divisor $\underline{\mathrm{Irr}}(\mathcal{M})$.)

(ii) Let E be the support of $\underline{\mathrm{Irr}}(\mathcal{M})$. By writing $|_E$ we mean the result of taking $\otimes_{\mathcal{O}_X} \mathcal{O}_E$. Then $\omega|_E$ does not depend on the choice of e and is a global section of $\Omega^1_{X/k}(\log D)|_E(\underline{\mathrm{Irr}}(\mathcal{M}))$. Write $\omega_\mathcal{M}$ for this global section.

(iii) We say that the triple $(X, U, \mathcal{M})$ is *clean* if, at any point x of E, the germ ω_x is a part of some basis of the stalk $\Omega^1_{X/k}(\log D)(\underline{\mathrm{Irr}}(\mathcal{M}))_x$.

(iv) We shall define the characteristic cycle $CC(\mathcal{M})$ of $j_*(\mathcal{M})$ on the logarithmic cotangent bundle $V(\Omega^1_{X/k}(\log D))$. Here $V(\ )$ denotes the vector bundle corresponding (covariantly) to a locally free sheaf of finite rank. Although characteristic cycles are usually defined on the cotangent bundle $V(\Omega^1_{X/k})$, the log-pole version is better suited to keep our analogy with Case I. Let $\mathcal{D}_{X/k,\log D} \subset \mathcal{D}_{X/k}$ be the sheaf of differential operators on X that preserve the defining ideal of each irreducible component of D. In the same way as in the case of $\mathcal{D}_{X/k}$-modules (cf. [47], Ch. 2, §8), our $CC(\mathcal{M})$ is defined naturally on $V(\Omega^1_{X/k}(\log D))$ as the characteristic cycle of the $\mathcal{D}_{X/k,\log D}$-module $j_*(\mathcal{M})$. If the triple $(X, U, \mathcal{M})$ is clean, we have

$$CC(\mathcal{M}) = [X] + \sum_{\mathfrak{p}} \mathrm{Irr}_{\mathfrak{p}}(\mathcal{M}) \cdot V(\mathrm{Image}(\varphi_{\mathfrak{p}})),$$

where $[X]$ is the 0-section of $V(\Omega^1_{X/k}(\log D))$, the sum is over the generic points $\mathfrak{p}$ of E, and $\varphi_\mathfrak{p}$ is the map "taking the product with $\omega_\mathcal{M}$":

$$\mathcal{O}_X(-\underline{\mathrm{Irr}}(\mathcal{M}))|_{\overline{\{\mathfrak{p}\}}} \to \Omega^1_{X/k}(\log D)|_{\overline{\{\mathfrak{p}\}}}$$

($\overline{\{\mathfrak{p}\}}$ is regarded as a reduced scheme). Note that, by cleanness, $\mathrm{Image}(\varphi_\mathfrak{p})$ is locally an $\mathcal{O}_{\overline{\{\mathfrak{p}\}}}$-direct summand. Being a subbundle of $V(\Omega^1_{X/k}(\log D)) \times_X \overline{\{\mathfrak{p}\}}$, $V(\mathrm{Image}(\varphi_\mathfrak{p}))$ is regarded as a cycle on $V(\Omega^1_{X/k}(\log D))$.

(v) Define $C(\mathcal{M}) \in CH_0(E)$ as an intersection

$$C(\mathcal{M}) = (-1)^{\dim(X)-1}[X] \cdot (CC(\mathcal{M}) - [X]).$$

Then the log-pole version of the formula of Dubson-Kashiwara is

$$\chi(U, \mathrm{DR}(\mathcal{M})) = (-1)^{\dim(X)} \deg([X] \cdot CC(\mathcal{M}))$$
$$= \chi(U) - \deg(C(\mathcal{M})),$$

where $\mathrm{DR}(\mathcal{M})$ is the de Rham complex $\mathcal{M} \otimes_{\mathcal{O}_U} \Omega_{U/k}$ associated with $\mathcal{M}$, and we define $\chi(U, \mathrm{DR}(\mathcal{M})) \stackrel{\mathrm{def}}{=} \sum_q (-1)^q \dim_k H^q(U, \mathrm{DR}(\mathcal{M}))$ ([32]). Our (4.2.4) is the analogue of this formula.

Now we consider Case I. First, we define the analogue of the logarithmic cotangent bundle (but only on E). Set

$$\Omega^1_X(\log D) = (\Omega^1_{X/\mathbb{Z}} \oplus (\mathcal{O}_X \otimes_\mathbb{Z} j_*\mathcal{O}_X^\times))/\mathcal{F},$$

where $\mathcal{F}$ is the subsheaf generated by local sections of the form $(da, 0) - (0, a \otimes a)$

($a \in \mathcal{O}_X \cap j_*\mathcal{O}_U^\times$). For sections $a \in \mathcal{O}_X$ and $b \in j_*\mathcal{O}_U^\times$, we write symbolically adb/b for the class $\in \Omega_X^1(\log D)$ of $(0, a \otimes b)$.

PROPOSITION 4.3.1. *The sheaf $\Omega_X^1(\log D)|_E$ is a locally free $\mathcal{O}_E$-module of finite rank, and its rank at a closed point x of E is equal to the dimension of the local ring $\mathcal{O}_{X,x}$.*

EXAMPLE. If X is of finite type over $\mathbb{Z}$ and of dimension n, then $\Omega_X^1(\log D)|_E$ has rank n. So Spec($\mathbb{Z}$) has, unexpectedly, "logarithmic cotangent bundles" of rank one (though they appear only on E and not on all of Spec($\mathbb{Z}$)). Indeed, if $X = \text{Spec}(\mathbb{Z})$ and $D = E = \{p\}$ with p a prime number, we have

$$\Omega_X^1(\log D)|_E \simeq \mathbb{F}_p; \quad \frac{dp}{p} \leftarrow 1.$$

(i) The divisor $\underline{\text{Sw}}(\rho)$. For each generic point $\mathfrak{p}$ of D, let $K_\mathfrak{p}$ be the fraction field of $\widehat{\mathcal{O}}_{X,\mathfrak{p}}$; it is a complete discrete valuation field. Given a representation $\rho : \pi_1(U) \to \mathbb{C}^\times$, we regard the representation $\rho_\mathfrak{p} : \text{Gal}((K_\mathfrak{p})^{\text{ab}}/K_\mathfrak{p}) \to \mathbb{C}^\times$ induced by ρ as a homomorphism into $\mathbb{Q}/\mathbb{Z}$, and hence an element of $H^1(K_\mathfrak{p})$, via the identification of the group of roots of unity in $\mathbb{C}^\times$ and $\mathbb{Q}/\mathbb{Z}$. Using the increasing filtration on $H^1(K_\mathfrak{p})$ mentioned at the end of §3.3, we define an integer

$$\text{Sw}_\mathfrak{p}(\rho) = \min\{n \geq 0; \rho_\mathfrak{p} \in \text{fil}_n H^1(K_\mathfrak{p})\},$$

and a divisor on X

$$\underline{\text{Sw}}(\rho) = \sum_\mathfrak{p} \text{Sw}_\mathfrak{p}(\rho) \cdot \mathfrak{p}.$$

(ii) In Case I, the analogue of the local section ω of $\Omega_{X/k}^1(\log D)(\underline{\text{Irr}}(\mathcal{M}))$ does not exist; nevertheless, the analogue ω_ρ of $\omega_\mathcal{M} = \omega|_E$ does exist. The section ω_ρ is characterized by the germs $\omega_{\rho,\mathfrak{p}}$ at the generic points $\mathfrak{p}$ of E as follows: put $i = \text{Sw}_\mathfrak{p}(\rho)$ and choose a prime element π of $K_\mathfrak{p}$. Using the map $\varphi_{\pi,i}^* : \text{gr}_i H^1(K_\mathfrak{p}) \to \Omega_{\kappa(\mathfrak{p})/\mathbb{Z}}^1 \oplus \kappa(\mathfrak{p})$ of (3.3.2), define two elements $\alpha \in \Omega_{\kappa(\mathfrak{p})/\mathbb{Z}}^1$ and $\beta \in \kappa(\mathfrak{p})$ by $(\alpha, \beta) = \varphi_{\pi,i}^*(\rho_\mathfrak{p})$. Then we have

$$\omega_{\rho,\mathfrak{p}} = (\alpha + \beta \frac{d\pi}{\pi}) \otimes \pi^{-i} \in \Omega_X^1(\log D)|_E(\underline{\text{Sw}}(\rho))_\mathfrak{p}.$$

(The right-hand side does not depend on the choice of π.)

(iii) We say that the triple (X, U, ρ) is *clean* if, for all $x \in E$, the germ $\omega_{\rho,x}$ is a part of some basis of the free $\mathcal{O}_{E,x}$-module $\Omega_X^1(\log D)|_E(\underline{\text{Sw}}(\rho))_x$.

EXAMPLE. Let k be a field of characteristic $p > 0$ and $X = \text{Spec}(k[T_1, T_2])$. Let D be the divisor $\{T_1 = 0\} \cup \{T_2 = 0\}$. Consider the character $\rho : \pi_1(U) \to \mathbb{C}^\times$ of order p that corresponds to the Artin-Schreier equation $\alpha^p - \alpha = g/T_1^m T_2^n$ ($m, n \geq 0$ with $p \nmid m$ or $p \nmid n$; $g \in k[T_1, T_2]$ with $T_1 \nmid g$ and $T_2 \nmid g$). Then we have

$$\underline{\text{Sw}}(\rho) = m\{T_1 = 0\} + n\{T_2 = 0\}, \quad \omega_\rho = d(\frac{g}{T_1^m T_2^n}),$$

and

$$(X, U, \rho) \text{ is clean} \Leftrightarrow g(0,0) \neq 0.$$

(Comparison: Consider Case II with k a field of characteristic 0, $X = \text{Spec}(k[T_1, T_2])$, and $D = \{T_1 = 0\} \cup \{T_2 = 0\}$. If $r = 1$, $\omega = d(g/T_1^m T_2^n)$ with $m, n \geq 0$, and $g \in k[T_1, T_2]$ with $T_1 \nmid g$, $T_2 \nmid g$, then we have

$$\underline{\mathrm{Irr}}(\mathscr{M}) = m\{T_1 = 0\} + n\{T_2 = 0\},$$

and

$(X, U, \mathscr{M})$ is clean $\Leftrightarrow g(0, 0) \neq 0$.)

(iv) It seems difficult to define in Case I an analogue of the object (iv) in Case II, unless the triple is clean. If it is clean, we define the analogue $CS(\rho)$ of $CC(\mathscr{M}) - [X]$ by

$$CS(\rho) = \sum_{\mathfrak{p}} \mathrm{Sw}_{\mathfrak{p}}(\rho) \cdot V(\mathrm{Image}(\varphi_{\mathfrak{p}}))$$

as a cycle on the vector bundle $V(\Omega_X^1(\log D)|_E)$ on E. Here the sum is over all generic points $\mathfrak{p}$ of E, and $\varphi_{\mathfrak{p}}$ is the map "taking the product with ω_ρ":

$$\mathscr{O}_X(-\underline{\mathrm{Sw}}(\rho))|_{\overline{\{\mathfrak{p}\}}} \to \Omega_X^1(\log D)|_{\overline{\{\mathfrak{p}\}}}.$$

(v) If (X, U, ρ) is clean, we define $C(\rho) \in CH_0(E)$ to be $(-1)^{\dim(E)}$ times the intersection of $CS(\rho)$ and the 0-section of $V(\Omega_X^1(\log D)|_E)$. When the triple is not clean, we can define $C(\rho)$ if $\dim(X) \leq 2$ by virtue of the following result.

THEOREM 4.3.2. *If* $\dim(X) \leq 2$, *then there exists a proper birational morphism* $g : X' \to X$ *such that g restricts to an isomorphism* $g^{-1}(U) \xrightarrow{\sim} U$, *such that $g^{-1}(D)$ viewed as a reduced scheme is a divisor with normal crossings on X', and such that $(X', g^{-1}(U), \rho)$ is clean.*

(The analogue in Case II of this theorem is proved for arbitrary dimension.)

If $\dim(X) \leq 2$, take an X' as in (4.3.2) and define $C(\rho)$ to be the image by g_* of the $C(\rho)$ for $(X', g^{-1}(U), \rho)$. This does not depend on the choice of X'.

In the above, we considered $C(\rho)$ for (X, U, ρ) only in the case where X is regular and $(X - U)_{\mathrm{red}}$ is a normal crossing divisor on X. But in general, if X is excellent with $\dim(X) \leq 2$ and U is regular, we can define $C(\rho)$ by reducing, via resolution of singularities, to the case of a normal crossing divisor (and by taking the direct image of the $C(\rho)$ of this case).

REMARK. In Case I (with $r = 1$), take a finite surjective morphism $Y \to X$, with Y normal, corresponding to the kernel of ρ. It is plausible that there is a strong relation between the two conditions that ρ be clean (this is a class-field-theoretic notion) and that the singularity of Y be not very bad. It would be fun to watch a wrestling match between Teiji Takagi (class field theory) and Heisuke Hironaka (resolution singularities) in the ring provided by our generalized class field theory.

Information on abelian extensions of a finitely generated field over a prime field is, in principle, all contained in its K-theoretic idele class group. The contents of §4 are fruits of an attempt to read off the aspect of the gr of the filtration on Milnor K-groups as we described in the Example at the end of §3.3, (ii), and to understand the drama of the storm brought by wild ramification on the ocean of differentials inherent in the idele class groups of global fields. I believe that if we listen sincerely to the words that K-groups tell us, we must be able, in the higher-dimensional case as well, to "tell perfectly what happens in abelian extensions as if it were happening on the palms of our hands" as declared in the Principal Spirit (1.0.1) of class field theory, including phenomena about singularities.

References

References related to generalizations of class field theory to higher-dimensional schemes:
1. S. Bloch, *Algebraic K-theory and class field theory for arithmetic surfaces*, Annals of Math. **114** (1981), 229–266.
2. J.-L. Brylinski, *Théorie du corps de classes de Kato et revêtements abéliens de surfaces*, Ann. Inst. Fourier **33-3** (1983), 23–38.
3. J.-L. Colliot-Thélène, J.-J. Sansuc, and C. Soulé, *Torsion dans le groupe de Chow de codimension deux*, Duke Math. J. **50** (1983), 763–801.
4. K. R. Coombes, *Local class field theory for curves*, Contemporary Math. **55, I** (1983), 117–134.
5. M. Gros, *Sur la partie p-primaire du groupe de Chow de codimension deux*, Communication in Algebra **13** (1985), 2407–2420.
6. Y. Ihara, *On a problem on some complete p-adic fields*, Kokyuroku, RIMS, Kyoto Univ. **41** (1968), 7–17. (Japanese)
7. K. Kato, *A generalization of local class field theory by using K-groups*. I, J. Fac. Sci. Univ. Tokyo, IA **26** (1979), 303–376.
   ____,II, ibid. **27** (1980), 603–683.
   ____,III, ibid. **29** (1982), 31–43.
8. K. Kato, *The existence theorem for higher local class field theory*, Inst. Hautes Études Sci. Publ. Math., preprint, 1980.
9. K. Kato, *Class field theory and algebraic K-theory*, Lecture Notes in Math., vol. 1016, Springer-Verlag, 1983.
10. K. Kato, *A Hasse principle for two dimensional global fields*, J. reine angew. Math. **366** (1986), 142–183.
11. K. Kato (with the collaboration of Takeshi Saito), *Vanishing cycles, ramification of valuations, and class field theory*, Duke Math. J. **55** (1987), 629–659.
12. K. Kato, *Class field theory, $\mathscr{D}$-modules, and ramification on higher dimensional schemes*. I, Amer. J. Math. **116** (1994), 757–784.
    ____,II, in preparation.
13. K. Kato and S. Saito, *Unramified class field theory of arithmetic surfaces*, Annals of Math. **118** (1983), 241–276.
14. K. Kato and S. Saito, *Two dimensional class field theory*, Advanced Studies in Pure Math. **2** (1983), 103–152.
15. K. Kato and S. Saito, *Global class field theory of arithmetic schemes*, Contemporary Math. **55, I** (1986), 255–331.
16. N. Katz and S. Lang, *Finiteness theorems in geometric class field theory*, l'Enseignement Mathématique **27** (1981), 285–314.
17. S. Lang, *Unramified class field theory over function fields in several variables*, Annals of Math. **64** (1956), 286–325.
18. V. G. Lomadze, *On the ramification theory of two dimensional local fields*, Math. U.S.S.R. Sbornik **37** (1980), 349–365.
19. A. N. Paršin, *Abelian coverings of arithmetic schemes*, Soviet Math. Dokl. **19, n°6** (1978), 1438–1442.
20. A. N. Paršin, *Local class field theory*, Proc. Steklov Inst. Math. **165** (1985), 157–185.
21. S. Saito, *Unramified class field theory for arithmetic schemes*, Annals of Math. **121** (1985), 251–281.
22. S. Saito, *Local class field theory for curves over a local field*, J. Number Theory **21** (1985), 44–80.
23. S. Saito, *Class field theory for two dimensional local rings*, Advanced Studies in Pure Math. **12** (1987), 343–373.
24. J.-P. Serre, *Groupes algébriques et corps de classes*, Hermann, Paris, 1959.

N.B. The idele class group we used in §2 of this article, which is a little different from that in [15], will be explained in detail in a joint paper (in preparation) with Shuji Saito.

The present article is titled "Generalization of class field theory"; however, I did not mention at all generalization in the direction of nonabelian theory.

Other References:
25. S. Bloch, K_2 and algebraic cycles, Annals of Math. **99** (1974), 349–376.
26. S. Bloch, Algebraic K-theory and crystalline cohomology, Inst. Hautes Études Sci. Publ. Math. **47** (1977), 187–268.
27. S. Bloch, Cycles on arithmetic schemes and Euler characteristics of curves, Proc. Sympos. Pure Math. **46, Part 2** (1987), 421–451.
28. S. Bloch and K. Kato, p-adic étale cohomology, Inst. Hautes Études Sci. Publ. Math. **63** (1986), 107–152.
29. R. Coleman and W. McCallum, Stable reduction of Fermat curves and Jacobi sum Hecke characters, J. reine angew. Math. **385** (1988), 41–101.
30. J.-L. Colliot-Thélène and J.-J. Sansuc, Chow groups of certain rational surfaces: A sequel to a paper of S. Bloch, Duke Math. J. **48** (1981), 421–447.
31. P. Deligne, La conjecture de Weil. II, Inst. Hautes Études Sci. Publ. Math. **52** (1980), 137–252.
32. A. Dubson, Formule pour l'indice des complexes constructibles et $\mathscr{D}$-modules holonomes, C. R. Acad. Sci. **209, Sér. A, 6** (1984), 113–116.
33. J. Graham, Continuous symbols on fields of formal power series, Lecture Notes in Math., vol. 342, Springer-Verlag, 1972, pp. 474–486.
34. O. Hyodo, Wild ramification in the imperfect residue field case, Advanced Studies in Pure Math. **12** (1987), 287–314.
35. O. Hyodo, A note on p-adic étale cohomology in the semi-stable reduction case, Invent. Math. **91** (1988), 543–557.
36. K. Kato, Milnor K-theory and the Chow group of zero cycles, Contemporary Math. **55, I** (1986), 241–253.
37. K. Kato, Swan conductors with differential values, Advanced Studies in Pure Math. **12** (1987), 315–342.
38. K. Kato, Swan conductors for characters of degree one in the imperfect residue field case, Contemporary Math. **83** (1989), 101–132.
39. K. Kato, Duality theories for the p-primary étale cohomology. I, Algebraic and Topological Theories (Kinosaki, 1984), Kinokuniya-shoten, Tokyo, 1985, pp. 127–148.
    ___,II, Compositio Math. **63** (1987), 259–270.
    ___,III, in preparation.
40. K. Kato, S. Saito, and T. Saito, Artin characters for algebraic surfaces, Amer. J. Math. **110** (1988), 49–76.
41. M. Kurihara, On two types of complete discrete valuation fields, Compositio Math. **63** (1987), 237–257.
42. M. Kurihara, Abelian extensions of absolutely unramified complete discrete valuation fields, Invent. Math. **93** (1988), 451–480.
43. G. Laumon, Caractéristique d'Euler-Poincaré des faisceaux constructibles sur une surface, Astérisque **101/102** (1983), 193–207.
44. Yu. I. Manin, Le groupe de Brauer-Grothendieck en géométrie diophantienne, Actes du Congrès Internat. Math., Nice, 1970, Gauthier-Villars, Paris, 1971, pp. 401–411.
45. J. Milnor, Algebraic K-theory and quadratic forms, Invent. Math. **9** (1970), 318–344.
46. H. Miki, On $\mathbb{Z}_p$-extensions of complete p-adic power series fields and function fields, J. Fac. Sci. Univ. Tokyo, IA **21** (1974), 377–393.
47. F. Pham, Singularités des systèmes différentiels de Gauss-Manin, Progress in Math., Birkhäuser, 1979.
48. D. Quillen, Higher algebraic K-theory. I, Lecture Notes in Math., vol. 341, Springer-Verlag, 1973, pp. 85–147.
49. S. Saito, Generalized fixed point formula for an algebraic surface and the theory of Swan representations of two dimensional local rings, Amer. J. Math. **109** (1987), 1009–1042.
50a. T. Saito, Bloch's 0-cycles and Artin characters of arithmetic surfaces, preprint.
50b. T. Saito, The Euler numbers of ℓ-adic sheaves of rank 1 in positive characteristic, Algebraic Geometry and Analytic Geometry, ICM-90 Satellite conference proceedings, Springer-Verlag, 1991, pp. 165–181.
51. J.-P. Serre, Sur la rationalité des représentations d'Artin, Annals of Math. **72** (1960), 406–420.
52a. J.-P. Serre, Sur les corps locaux à corps résiduel algébriquement clos, Bull. Soc. Math. France **89** (1961), 105–154.
52b. M. Hazewinkel, Corps de classes local, Appendix to: Groupes Algébriques (by M. Demazure and P.Gabriel), North-Holland Publ., 1970, pp. 648–674.

53. J.-P. Serre, *Zeta and L-functions*, Arithmetic Algebraic Geometry, Harper and Row, New York, 1963, pp. 82–92.
54. C. Soulé, *Opérations en K-théorie algébrique*, Canadian J. Math. **37** (1985), 488–550.
55. T. Takagi, *Algebraic Number Theory*, Iwanami-Shoten, Tokyo, 1971 (2nd ed.). (Japanese)

Translator's reference:
90. K. Kato, *Generalized class field theory*, Proc. ICM-90, Kyoto, Springer-Verlag, 1991, pp. 419–428.

The translator thanks Kazuya Kato and Wayne Raskind for their correcting the text of the first draft of this translation. Thanks are also due to Wayne Raskind who drew the translator's attention to some of the papers by Fesenko on higher local class field theory, especially to the following ones:

[F1] I. Fesenko, *On class field theory of multidimensional local fields of positive characteristic*, Algebraic K-Theory, Advances in Soviet Math., vol. 4, Amer. Math. Soc., Providence, RI, 1991.
[F2] I. Fesenko, *A multidimensional local class field theory II*, St. Petersburg Math. J. **3** (1992), 1101–1126.
[F3] I. Fesenko, *Class field theory of multidimensional local fields of characteristic zero with residue field of positive characteristic*, St. Petersburg Math. J. **3** (1992), 649–678.
[F4] I. Fesenko, *Local fields, local class field theory, higher local class field theory via algebraic K-theory*, St. Petersburg Math. J. **4** (1993), 403–438.

A more extensive bibliography including some references published after the publication of the original article of Kato can be found in
[R] W. Raskind, *Abelian class field theory of arithmetic schemes*, to appear in Proc. Sympos. Pure Math.

DEPARTMENT OF MATHEMATICS, UNIVERSITY OF TOKYO, HONG, BUNKYO, TOKYO 113, JAPAN

Translated by YUICHIRO TAGUCHI

Recent Topics on Open Algebraic Surfaces

Masayoshi Miyanishi

§1. Introduction

Let us consider a nonsingular curve C of degree d on the projective plane $\mathbf{P}^2$. Let $X = \mathbf{P}^2 - C$ be the complement of C. For the pair $(\mathbf{P}^2, C)$, we have the following exact sequence of integral cohomologies:

$$\begin{aligned}
0 &\to H^0(\mathbf{P}^2, C) \to H^0(\mathbf{P}^2) \to H^0(C) \\
&\to H^1(\mathbf{P}^2, C) \to H^1(\mathbf{P}^2) \to H^1(C) \\
&\to H^2(\mathbf{P}^2, C) \to H^2(\mathbf{P}^2) \to H^2(C) \\
&\to H^3(\mathbf{P}^2, C) \to H^3(\mathbf{P}^2) \to 0 \\
&\to H^4(\mathbf{P}^2, C) \to H^4(\mathbf{P}^2) \to 0.
\end{aligned}$$

Noting that $H^i(\mathbf{P}^2) \cong \mathbf{Z}$ if $i = 0, 2, 4$ and $H^i(\mathbf{P}^2) = (0)$ if $i = 1, 3$ and that $H^0(C) \cong H^2(C) \cong \mathbf{Z}$ and $H^1(C) \cong \mathbf{Z}^{2g}$, we deduce from the above exact sequence that

$$H^0(\mathbf{P}^2, C) \cong H^1(\mathbf{P}^2, C) = (0), \qquad H^2(\mathbf{P}^2, C) \cong \mathbf{Z}^{2g},$$
$$H^3(\mathbf{P}^2, C) \cong \mathbf{Z}/d\mathbf{Z}, \qquad H^4(\mathbf{P}^2, C) \cong \mathbf{Z},$$

where $g = \frac{1}{2}(d-1)(d-2)$ is the genus of the curve C. Then, by virtue of the Lefschetz duality,

$$H_i(X) \cong H^{4-i}(\mathbf{P}^2, C), \quad 0 \le i \le 4,$$

we obtain

$$H_0(X) \cong \mathbf{Z}, \quad H_1(X) \cong \mathbf{Z}/d\mathbf{Z}, \quad H_2(X) \cong \mathbf{Z}^{2g}, \quad H_3(X) \cong H_4(X) \cong (0).$$

Moreover, it is known by Zariski's theorem (cf. [73, 20, 4]) that the complement of a curve C on $\mathbf{P}^2$ has an abelian fundamental group provided the irreducible components of C meet each other transversally; we allow the irreducible components to have *nodal* singularities. Hence $\pi_1(X) \cong H_1(X) \cong \mathbf{Z}/d\mathbf{Z}$.

If we allow the curve C to have singularities or to split into several components, the homology groups change accordingly. We exhibit the change in the case $d = 3$,

---

1991 *Mathematics Subject Classification*. Primary 14J29, 30F20, 57N05.

This article originally appeared in Japanese in Sūgaku **46** (3) (1994), 243–257.

i.e., the case of degeneration of an elliptic curve. Since $H_0(X) \cong \mathbf{Z}$ and $H_3(X) \cong H_4(X) \cong (0)$ for each case of degeneration, we only trace the change of $H_1(X)$ and $H_2(X)$ in the following table:

type of degeneration	smooth	nodal	cuspidal	3 lines	3 confluent lines	1 double line plus 1 line	1 triple line
$H_1(X)$	$\mathbf{Z}/3\mathbf{Z}$	$\mathbf{Z}/3\mathbf{Z}$	$\mathbf{Z}/3\mathbf{Z}$	$\mathbf{Z}^2$	$\mathbf{Z}^2$	$\mathbf{Z}$	0
$H_2(X)$	$\mathbf{Z}^2$	$\mathbf{Z}$	0	$\mathbf{Z}$	0	0	0

Here we note that X is a nonsingular affine complex surface. So, $H_i(X) \cong (0)$ for $i > \dim X$ by virtue of a theorem of Andreotti and Frankel [1]. See also [39].

A nonsingular affine complex surface X is called a *homology plane* (or **Q**-*homology plane*) if $H_i(X; \mathbf{Z}) \cong (0)$ (or $H_i(X; \mathbf{Q}) \cong (0)$) for $i = 1, 2$. So, the above table shows that $X = \mathbf{P}^2 - C$ is a homology plane only if C is a triple line and that X is a **Q**-homology plane if C is a cuspidal cubic or a triple line. If C is a triple line, then X is isomorphic to the affine plane $\mathbf{A}^2$; we denote the affine space of dimension n by $\mathbf{A}^n$ or $\mathbf{C}^n$.

Van de Ven [69] first considered a complex homology 2-cell, which is, by definition, a smooth complex analytic surface with the vanishing reduced integral homologies, and he proved that a complex homology 2-cell is algebraic, i.e., it is embeddable into a nonsingular projective surface as a Zariski open set. Thus, it is a homology plane. Gurjar and Shastri [26] proved that a homology plane is rational, i.e., the field of algebraic functions on a homology plane is a purely transcendental extension of dimension two over the complex field **C**. However, the rationality of the **Q**-homology plane is not known.

One may ask the following questions related to homology planes:
(1) Is a homology plane isomorphic to $\mathbf{A}^2$, especially when it is topologically contractible?
(2) If not, describe the structures of homology planes and classify them.

The first question was probably asked in connection with the following problem.

CANCELLATION PROBLEM. Let X and Y be complex algebraic surfaces. Assume that $X \times \mathbf{C}$ is biregular (or biholomorphic) to $Y \times \mathbf{C}$. Is X isomorphic to Y? (If $Y \cong A^2$, then X is topologically contractible.)

As we observe below, the answer is affirmative if $Y = \mathbf{A}^2$ by Fujita [17], Miyanishi and Sugie [47] (see also Kambayashi [30]), or if X and Y have nonnegative (logarithmic) Kodaira dimension by Fujita and Iitaka [21], but the answer is negative in general as shown by Danielewski [3] and Fieseler [80].

The second question is our main theme. Namely, how do we describe the topological, algebro-geometric, and analytic structures of homology planes, **Q**-homology planes, or non-complete algebraic surfaces in general?

In the present article, we assume that the ground field is the complex field **C**.

§2. Ramanujam surface

It was C. P. Ramanujam [59] who first constructed a contractible homology plane that is not biregular to $\mathbf{A}^2$. He found an affine surface, which is now called a *Ramanujam surface*, in his trial to find a counterexample to the cancellation problem (when $Y = \mathbf{A}^2$). We shall discuss his construction.

On the projective plane $\mathbf{P}^2$, consider a cubic curve C_1 with a cusp R and an irreducible conic C_2 meeting each other in two points P, Q, other than R, with respective multiplicities 5, 1. For example, we may take the curves C_1, C_2 defined by the following equations with respect to homogeneous coordinates (X_0, X_1, X_2),

$$C_1: X_2^2 X_0 = X_1^3,$$
$$C_2: X_0^2 - \tfrac{3}{5} X_0 X_1 - \tfrac{9}{125} X_1^2 - \tfrac{8}{25} X_0 X_2 - \tfrac{24}{3125} X_1 X_2 - \tfrac{1}{3125} X_2^2 = 0.$$

Three points P, Q, R are then given as

$$P = (-\tfrac{1}{125}, -\tfrac{1}{5}, 1), \qquad Q = (1, 1, 1), \qquad R = (1, 0, 0).$$

Let $\sigma: F \to \mathbf{P}^2$ be the blowing-up with center Q, let $E = \sigma^{-1}(Q)$ be the exceptional curve and let C_i' ($i = 1, 2$) be the proper transform of C_i, i.e., the closure of $C_i - \{Q\}$ in F. Here we recall that σ induces an isomorphism $F - E \xrightarrow{\sim} \mathbf{P}^2 - \{Q\}$ and the points on E correspond bijectively with the tangential directions of the lines on $\mathbf{P}^2$ passing through Q. Set $X := F - C_1' \cup C_2'$, which we call a *Ramanujam surface*. The surface X enjoys the properties described in the following theorem.

THEOREM 1. *Let X be a Ramanujam surface. Then the following assertions hold*:
(1) X *is a topologically contractible homology plane*,
(2) X *is not biregular to $\mathbf{A}^2$*,
(3) $X \times \mathbf{C}$ *is diffeomorphic to $\mathbf{C}^3$, but not biregular to $\mathbf{C}^3$*.

The surface F has a $\mathbf{P}^1$-bundle structure $f: F \to E$, whose fibers are the proper transforms by σ of the lines through Q, and C_2' is a cross section of f because a line through Q meets C_2 in another point by Bézout's theorem. Thus, f induces an $\mathbf{A}^1$-bundle structure $f: F - C_2' \to E$.

We may (and shall) choose an inhomogeneous coordinate z on E so that $f(R) = 0$ and $f(P) = \infty$. Then $G := F - C_2' - f^{-1}(\infty)$ is an $\mathbf{A}^1$-bundle over $\mathbf{A}^1 \cong E - \{\infty\}$. Since an $\mathbf{A}^1$-bundle over $\mathbf{A}^1$ is trivial, we may find a fiber coordinate w so that G is isomorphic to the (z, w)-plane $\mathbf{A}^2$, where f is given by the projection $(z, w) \mapsto z$. Since $f: G \cap C_1' \to \mathbf{A}^1$ is a finite covering of degree two by the restriction of f onto $G \cap C_1'$, the curve $G \cap C_1'$ is defined by an equation

$$w^2 + a(z)w - p(z) = 0.$$

By a change of coordinates $(z, w) \mapsto (z, w + \tfrac{1}{2} a(z))$, we may assume $a(z) = 0$.

It can be shown that there is a unique line ℓ through Q, other than the line $\overline{RQ}$, that is tangential to C_1 at a point, say A. For the above example, it is given by

$$X_0 = \tfrac{3}{4} X_1 + \tfrac{1}{4} X_2, \qquad A = (\tfrac{1}{4}, 1, -2).$$

We may choose the coordinate z so that the proper transform ℓ' of ℓ is defined by $z = 1$. Then the defining equation of $G \cap C_1'$ is written as

$$w^2 = p(z) = z^m (z - 1)^n,$$

where we have $m = 3$ and $n = 1$ because $G \cap C_1' \cap f^{-1}(0)$ is a cusp and $G \cap C_1' \cap f^{-1}(1)$ is a smooth point. Namely, $p(z) = z^3(z-1)$.

Moreover, the $\mathbf{A}^1$-bundle $f \colon G \to \mathbf{A}^1$ induces a locally trivial fiber space

$$f' \colon G' \to \mathbf{A}^1 - \{0, 1\}$$

whose fibers are isomorphic to $\mathbf{A}^1 - \{0, 1\}$, where $G' := G - C_1' - f^{-1}(0) - f^{-1}(1)$.

The homotopy exact sequence for the fiber space f' yields an exact sequence

$$\pi_1(f'^{-1}(\lambda)) \to \pi_1(G') \to \pi_1(E - \{0, 1, \infty\}) \to 0,$$

where $f'^{-1}(\lambda)$ is a general fiber of f', and the natural homomorphism $\pi_1(G') \to \pi_1(G - C_1')$ is a surjection because G' is a Zariski open set of $G - C_1'$. Hence $\pi_1(G - C_1')$ is generated by $\pi_1(f'^{-1}(\lambda))$ and the classes of loops η_0, η_1 around the fibers $f'^{-1}(0)$ and $f'^{-1}(1)$ if the base point is chosen suitably in between these two fibers. However, since the loops η_0, η_1 are contractible in $G - C_1'$, we know that $\pi_1(G - C_1')$ is generated by two loops in $f'^{-1}(\lambda)$ around the two points in $C_1' \cap f'^{-1}(\lambda)$.

Let $A = C_1' \cap f'^{-1}(1)$, which is the point where the line ℓ touches the curve C_1. Near the point A, we can find a system of local coordinates (z', w') such that the curve C_1' is defined near A by $z' = 0$. Let U be an open neighborhood of A,

$$U = \{(z', w') \mid |z'| < 1, |w'| < 1\}.$$

Then $U - C_1'$ is homeomorphic to $\Delta^* \times \Delta$, where Δ is the unit disk and $\Delta^* = \Delta - \{0\}$, and $\pi_1(U - C_1') \to \pi_1(G - C_1')$ is surjective. Since $\pi_1(U - C_1') \cong \pi_1(\Delta^*) \cong \mathbf{Z}$, we know that $\pi_1(G - C_1')$ is abelian.

Let $C' = C_1' \cup C_2'$ be the union of the curves C_1', C_2' meeting at a single point P on F. Concerning the integral homologies of F and C', we know that

$$H_1(C') = (0), \quad H_2(C') = \text{a free abelian group generated by the}$$
$$\text{classes } [C_1'] \text{ and } [C_2'],$$
$$H_1(F) = (0), \quad H_2(F) = \text{a free abelian group generated by the}$$
$$\text{class } [E] \text{ and the class } [H],$$

where $[H]$ is the class of the inverse image of a hyperplane H on $\mathbf{P}^2$, and where the canonical homomorphism $H_2(C') \to H_2(F)$ maps $[C_1']$ and $[C_2']$ to $3[H] - [E]$ and $2[H] - [E]$, respectively. The homology exact sequence for the pair (F, C') implies

$$H_i(F, C') = (0) \quad \text{for } 0 \le i \le 3.$$

By the universal coefficient theorem and Poincaré's duality, we know that

$$H_i(X) \cong H^{4-i}(F, C') = (0) \quad \text{for } 1 \le i \le 4.$$

Thus we have proved the following assertions:
(1) X is a homology plane,
(2) $\pi_1(X) = (0)$,
(3) hence X is contractible by a theorem of J. H. C. Whitehead.

Next, we shall show that X is not biregular to $\mathbf{A}^2$. For this purpose, we shall construct a nonsingular projective surface $\overline{X}$ satisfying the following conditions:
  (i) $\overline{X}$ contains X as a Zariski open set,
  (ii) the complement $D := \overline{X} - X$ consists of nonsingular irreducible curves such that no three of them meet in a point and any two of them meet each other

transversally if they meet in a point; we then say that D is a divisor with *simple normal crossings*.

Such a surface $\overline{X}$, called a *normal completion* (or a *normal compactification*), is obtained by applying successively the blow-ups with centers at the points P, R and the points appearing on the exceptional curves where the proper transforms of the curves C_1, C_2 still have singular points or meet each other not in a "normal crossings" way together with the exceptional curves.

We describe the effect of blow-ups by writing down the configuration of exceptional curves and the proper transforms of C_1, C_2, each exceptional curve being assigned the self-intersection number, or by expressing the configuration of those curves and their intersections by the weighted dual graph in which each irreducible curve corresponds to a vertex and two vertices are linked by an edge if the corresponding curves meet each other.

In our case, the surface $\overline{X}$ is obtained from F by applying altogether eight blow-ups, the configuration is given by Figure 1, and the weighted dual graph is given by Figure 2.

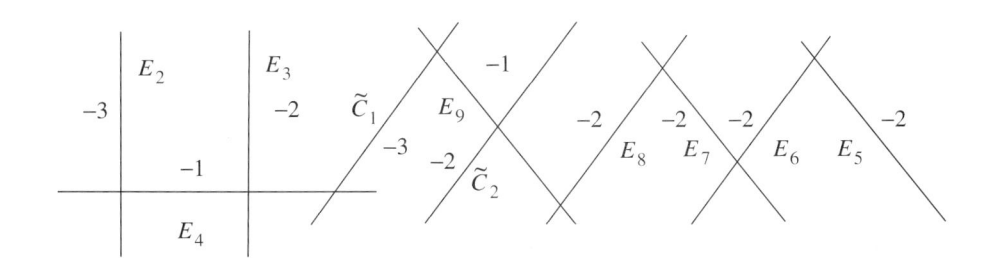

FIGURE 1

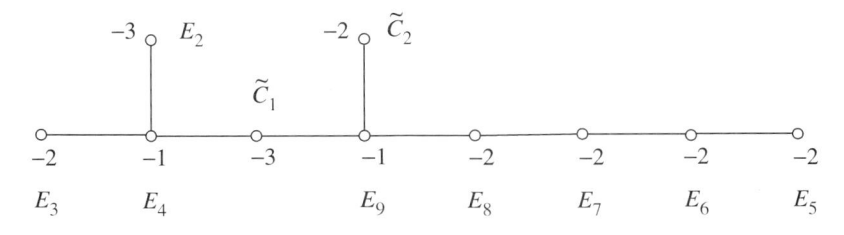

FIGURE 2

All the curves concerned are nonsingular rational curves.

Let $D = \overline{X} - X$. Then $D = \widetilde{C}_1 + \widetilde{C}_2 + E_2 + \cdots + E_9$, where the union of irreducible components is denoted additively to signify that D is considered a divisor on $\overline{X}$. D is called the *boundary* (or the *divisor*) *at infinity*. With respect to a suitable Riemannian metric on $\overline{X}$ and for $0 < \delta \ll 1$, let $\overline{V}_\delta$ be the set of points whose distance from D is $\leq \delta$, let S_δ be its boundary, and let $V_\delta = \overline{V}_\delta - S_\delta$. Then V_δ is a tubular neighborhood of D, D is a strong deformation retract of $\overline{V}_\delta$, and $\overline{V}_\delta - D$ is homotopy equivalent to S_δ. Let $M = \overline{X} - V_\delta$. Then M is a compact 4-manifold with boundary $\partial M = S_\delta$ and M is contractible since $X = \overline{X} - D$ is contractible.

Now, looking at the integral cohomology exact sequence for a pair $(\overline{V}_\delta, D)$ and using the Poincaré and Lefschetz dualities, we obtain

$$H_0(S_\delta) \cong \mathbf{Z}, \qquad H_2(S_\delta) \cong (0), \qquad H_3(S_\delta) \cong \mathbf{Z}$$

and an exact sequence

$$0 \to H_2(D) \xrightarrow{\gamma} H^2(D) \to H_1(S_\delta) \to 0.$$

Here the homomorphism γ is given by assigning to the class [A] the mapping $\varphi_{[A]}$ defined by $\varphi_{[A]}([B]) = (A, B)$, where $[A], [B]$ are the classes in $H_2(D)$ represented by irreducible components A, B of D, and (A, B) is the intersection number. Hence the order $|H_1(S_\delta)|$ is the absolute value of the determinant of the intersection matrix of D, which turns out to be 1. Hence $H_1(S_\delta) = (0)$. This implies that S_δ is a homology sphere, i.e., S_δ has the same homology groups as the 3-sphere S^3. However, S_δ is not a homotopy sphere.

In order to see this, consider the fundamental group $\pi_1(S_\delta)$, which is, in fact, independent of the choice of $\overline{V}_\delta$ (or δ) and is called the *fundamental group at infinity* of X, with the notation $\pi_1^\infty(X)$.

Ramanujam [59] proved the following theorem.

THEOREM 2. *Let X be a topologically contractible nonsingular complex affine algebraic surface. Then X is biregular to $\mathbf{A}^2$ if and only if $\pi_1^\infty(X) = (0)$.*

So, if we show that X is not biregular to $\mathbf{A}^2$, we know that S_δ is not a homotopy sphere. We can again use the result of Ramanujam [59] that the dual graph of the divisor at infinity for a minimal normal compactification of $\mathbf{A}^2$ is a linear chain. Here a compactification is called *minimal* if we cannot contract algebraically any irreducible component in the boundary at infinity (to a smooth point) without violating the property that the boundary divisor at infinity is a divisor of simple normal crossings. Then Morrow [56] classified the boundary dual graphs of minimal normal compactifications of $\mathbf{A}^2$.

In view of Figure 2, we know that $\overline{X}$ is a minimal normal compactification of X, that the dual graph of the boundary at infinity is not a linear chain, and that X is therefore not biregular to $\mathbf{A}^2$. Hence, S_δ is not a homotopy sphere.

Now, set $M' = M \times [0, 1] \times [0, 1]$ and

$$\partial M' = \partial M \times [0, 1] \times [0, 1] \cup M \times \{0, 1\} \times [0, 1] \cup M \times [0, 1] \times \{0, 1\}.$$

Then M' is contractible and $\partial M'$ is a homotopy sphere. In fact, $\partial M'$ is a homology sphere, and $\pi_1(\partial M') = (0)$ since $\pi_1(M) = (0)$. Hence $\partial M'$ is a homotopy sphere by a theorem of Hurewitz. Noting that $\dim_\mathbf{R} M' = 6$ and $\dim_\mathbf{R} \partial M' = 5$, we can make use of the generalized Poincaré conjecture in dimension ≥ 5 (see Smale [64]) to show that M' and $\partial M'$ are diffeomorphic to the disk D^6 and the 5-sphere S^5, respectively. Since δ is variable, we can conclude that $X \times \mathbf{R}^2$ is diffeomorphic to $\mathbf{R}^6$ (use Palais [57]). Thus we have completed the proof of Theorem 1.

COROLLARY 3. *The Euclidean 6-space $\mathbf{R}^6$ has an exotic complex structure $X \times \mathbf{C}$, i.e., a complex structure not biregular to $\mathbf{C}^3$; $X \times \mathbf{C}$ is not biholomorphic to $\mathbf{C}^3$ by Zaidenberg [71].*

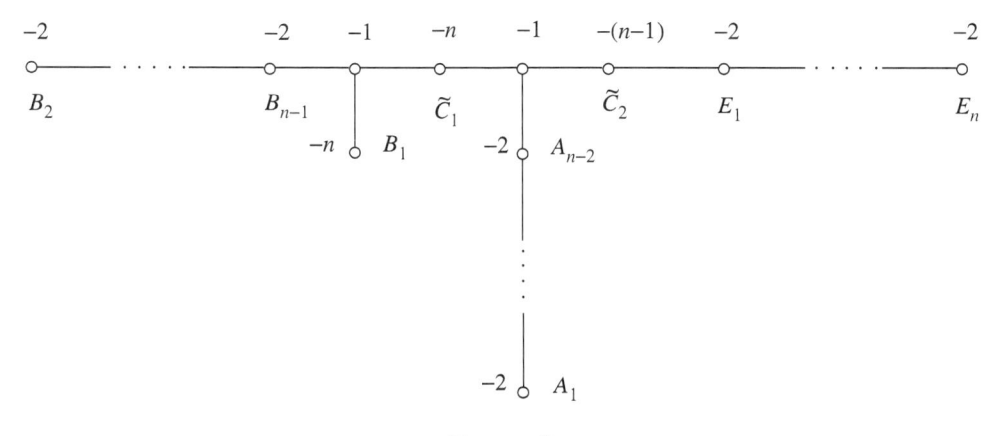

FIGURE 3

We can obtain a similar kind of nonsingular contractible affine algebraic surfaces X_n ($n \geq 3$) from the projective plane curves (see [66]):

$$C_1: X_2^{n-1} X_0 = X_1^2 \left(X_1^{n-2} - \sum_{i=2}^{n-2} (i-1)_n C_i X_1^{n-i-2} X_2^i \right)$$

and

$$C_2: X_0 = (n^2 - 2n)X_1 + (n-1)X_2$$

which meet in two points $P = (-(n^2 - 3n + 1), -1, 1)$ and $Q = ((n-1)^3, (n-1), 1)$; two curves touch in P with order of contact $(n-1)$ and in Q transversally. Moreover, C_1 has a singular point $R = (1, 0, 0)$. Let $\sigma: \widetilde{F} \to \mathbf{P}^2$ be a composite of blow-ups with center at Q and Q_i ($1 \leq i \leq n$), where Q_1 is the intersection point of the proper transform of C_1 and the exceptional curve of the blow-up with center Q, and inductively, Q_i is the intersection point of the proper transform of C_1 with the exceptional curve of the blow-up with center Q_{i-1}; we say that Q_1 is the *infinitely near point* of Q of *first order* lying on C_1 and that Q_i is the infinitely near point of Q of order i lying on C_1. Let E_{n+1} be the exceptional curve of the blow-up with center Q_n. Let $X_n = \widetilde{F} - (\sigma^{-1}(C_1 \cup C_2) - E_{n+1})$. Then X_n is the required surface and has the divisor at infinity whose weighted dual graph is given in Figure 3 and whose intersection matrix is unimodular.

Thus, as in the case of the Ramanujam surface, $X_n \times \mathbf{C}$ gives rise to an exotic complex structure on $\mathbf{C}^3$. So, $\mathbf{R}^6$ has infinitely many, mutually nonbiholomorphic, complex structures.[1] In fact, $X_n \times \mathbf{C} \cong X_m \times \mathbf{C}$ implies $X_n \cong X_m$ by a theorem of Fujita and Iitaka [21] and by a theorem of Zaidenberg [71] because both X_n and X_m have Kodaira dimension two (see §3 below).

§3. Kodaira dimension

Let X be a nonsingular algebraic surface. We say that X is an *open* (or synonymously, *non-complete*) algebraic surface if X is not complete (in fact, not projective).

---

[1] (Added to the English translation) Denote $\pi_1^\infty(X)/[\pi_1^\infty(X), \pi_1^\infty(X)]$ by $H_1^\infty(X; \mathbf{Z})$, where $[\pi_1^\infty(X), \pi_1^\infty(X)]$ is the commutator group. Then, for $n \geq 3$, $H_1^\infty(X_n; \mathbf{Z}) \cong \mathbf{Z} \oplus \mathbf{Z}/(n-1)\mathbf{Z}$. Hence X_n is not homeomorphic to X_m if $n \neq m$.

In the subsequent arguments, we consider as X an affine surface or $W - \text{Sing}(W)$, where W is a projective singular surface and $\text{Sing}(W)$ is the singular locus of W.

We can embed an open algebraic surface X into a nonsingular projective surface V so that the complement $D = V - X$ is a divisor with simple normal crossings. A pair (V, D) is a *normal completion* of X, and D is the *boundary at infinity* of X. If $C_1, \ldots, C_r$ are irreducible components of D, we write $D = C_1 + \cdots + C_r$ as a divisor. If X is affine, then there is a closed embedding $V \hookrightarrow \mathbf{P}^N$ such that a suitable hyperplane section of V has the same support as D; in other words, D supports an ample divisor of V. In particular, D is connected.

Let K_V be the canonical divisor of V; the associated sheaf $\mathscr{O}(K_V)$ is the sheaf of germs of regular differential 2-forms on V.

For an integer $n > 0$, set $h^0(n(D + K_V)) = \dim H^0(V, \mathscr{O}(D + K_V)^{\otimes n})$. An element $\varphi \in H^0(V, \mathscr{O}(D + K_V)^{\otimes n})$ is expressed locally at a point $P \in V$ as follows in terms of local coordinates $\{u, v\}$,

$$\varphi = \varphi_P (du \wedge dv)^n,$$

where φ_P is a rational function on V admitting poles of order at most n along the divisor D.

Set $\mathbf{N}(D, V) = \{n > 0; h^0(n(D + K_V)) \neq 0\}$. We define the $(D + K_V)$-dimension κ by

$$\kappa = \begin{cases} -\infty & \text{if } \mathbf{N}(D, V) = \varnothing, \\ \sup_{n \in \mathbf{N}(D,V)} \dim \Phi_n(V) & \text{if } \mathbf{N}(D, V) \neq \varnothing \end{cases}$$

where $\Phi_n \colon V \to \mathbf{P}^N$ is a rational mapping defined by $P \mapsto (\varphi_0(P), \ldots, \varphi_N(P))$ for a linearly independent basis $\{\varphi_0, \ldots, \varphi_N\}$ of $H^0(V, \mathscr{O}(D + K_V)^{\otimes n})$ with $N + 1 = h^0(n(D + K_V))$ and $\Phi_n(V)$ is the image of V by Φ_n. (See also Sakai [62].) By Iitaka [29], κ is independent of the choice of embeddings $X \hookrightarrow V$ as above. We therefore call κ the (logarithmic) *Kodaira dimension* of X and denote it by $\kappa(X)$. It takes one of the values $-\infty, 0, 1, 2$.

As in the complete case, the classification of open algebraic surfaces proceeds with the values of $\kappa(X)$. In what follows, we sketch the classification and state some of the noteworthy results on the structure of X that are necessary for the understanding of this article. For more precise treatments, we refer to Fujita [19], Kawamata [33], Miyanishi [43], Miyanishi and Tsunoda [50], Tsunoda [67], and Tsunoda and Miyanishi [68].

First of all, we shall consider the case $\kappa(X) = -\infty$.

THEOREM 4. *Let X be an open algebraic surface with $\kappa(X) = -\infty$. Suppose that the boundary at infinity is connected. Then there exists a morphism $\rho \colon X \to C$ from X to a nonsingular algebraic curve C whose general fibers are isomorphic to $\mathbf{A}^1$ or $\mathbf{P}^1$. (See Fujita [17], Miyanishi and Sugie [47], Sugie [65], and Russell [61].)*

We say that X has an $\mathbf{A}^1$-*fibration* if X has a morphism $\rho \colon X \to C$ as in Theorem 4 and that X is *affine-ruled* if X has an $\mathbf{A}^1$-fibration. If X is affine-ruled, then $\kappa(X) = -\infty$. We say that X has an $\mathbf{A}^1_*$-fibration (or $\mathbf{C}^*$-fibration) if there is a morphism $\rho \colon X \to C$ from X onto a nonsingular algebraic curve C whose general fibers are isomorphic to $\mathbf{A}^1_* := \mathbf{A}^1 - \{\text{one point}\}$. Then the function field $\mathbf{C}(X)$ is a finitely-generated extension of the function field $\mathbf{C}(C)$ of transcendence degree 1. We say that the $\mathbf{A}^1_*$-fibration $\rho \colon X \to C$ is *untwisted* (or *twisted*, resp.) if $\mathbf{C}(X)$ is a purely transcendental extension of $\mathbf{C}(C)$ (otherwise, resp.). A typical example of an open algebraic surface with such an (untwisted) $\mathbf{C}^*$-fibration is the quotient variety $\mathbf{C}^2/G$, where G is a finite subgroup of $\mathrm{GL}(2,\mathbf{C})$. The quotient variety $\mathbf{C}^2/G$ has the $\mathbf{C}^*$-action induced by the action of $\mathbf{C}^*$ on $\mathbf{C}^2$ via the center of $\mathrm{GL}(2,\mathbf{C})$, and $\mathbf{C}^2/G$ has the unique singular point P, which is the image of the point of origin of $\mathbf{C}^2$. Set $T := \mathbf{C}^2/G - \{P\}$. Then the surface T has the induced $\mathbf{C}^*$-fibration $\rho \colon T \to \mathbf{P}^1$. We call T a *Platonic $\mathbf{C}^*$-fiber space*. For the following result, we refer to Miyanishi and Tsunoda [50, 51] and Miyanishi [45].

THEOREM 5. *Let X be an open algebraic surface with $\kappa(X) = -\infty$. Suppose that the boundary at infinity $D = C_1 + \cdots + C_r$ of X is not connected. Then there exist a Zariski open set U of X and a proper birational morphism $\varphi \colon U \to T'$ onto an open set T' of an algebraic surface T satisfying the following conditions.*
  (1) *Either $U = X$ or $X - U$ has pure dimension one.*
  (2) $\dim(T - T') \leq 0$.
  (3) *Suppose that the intersection matrix $((C_i, C_j))_{1 \leq i,j \leq r}$ is not negative-definite. Then T is a Platonic $\mathbf{C}^*$-fiber space.*
  (4) *Suppose that the intersection matrix $((C_i, C_j))_{1 \leq i,j \leq r}$ is negative-definite. Then $T = W - \mathrm{Sing}(W)$, where W is a log del Pezzo surface of rank 1, i.e., W is a normal projective surface with only quotient singularities such that $-K_W$ is an ample $\mathbf{Q}$-Cartier divisor and $\dim_{\mathbf{Q}} \mathrm{Pic}(W) \otimes_{\mathbf{Z}} \mathbf{Q} = 1$, and $\mathrm{Sing}(W)$ is the singular locus of W.*

An application of Theorem 4 is the following characterization of the affine plane (cf. Miyanishi [43]).

THEOREM 6. *Let X be a nonsingular affine surface. Then X is isomorphic to the affine plane if and only if the following three conditions are satisfied*:
  (i) *X has an $\mathbf{A}^1$-fibration;*
  (ii) $\Gamma(\mathscr{O}_X)^* = \mathbf{C}^*$;
  (iii) $\Gamma(\mathscr{O}_X)$ *is a UFD,*
where $\Gamma(\mathscr{O}_X)$ is the ring of regular functions on X and $\Gamma(\mathscr{O}_X)^*$ is the multiplicative group of invertible elements of $\Gamma(\mathscr{O}_X)$.

Since the three conditions on X in Theorem 6 follow from the hypothesis $X \times \mathbf{A}^1 \cong \mathbf{A}^3$ in the cancellation problem, we know that $X \cong \mathbf{A}^2$ if $X \times \mathbf{A}^1 \cong \mathbf{A}^3$. Gurjar and Shastri [25] obtained the following topological characterization of $\mathbf{C}^2/G$.

THEOREM 7. *Let X be a normal affine surface. Then X is isomorphic to $\mathbf{C}^2/G$ if and only if X is topologically contractible and $\pi_1^\infty(X)$ is a finite group.*

A further application of Theorems 4 and 5 is the following result (cf. Miyanishi [45]).

THEOREM 8. *Let A be a normal subalgebra of a polynomial ring $\mathbf{C}[x_1, x_2]$ such that $\mathbf{C}[x_1, x_2]$ is a finitely-generated A-module and let $X = \mathrm{Spec}\, A$. Then $X \cong \mathbf{A}^2$ if X is nonsingular, and $X \cong \mathbf{C}^2/G$ if X is singular. Furthermore, if X is singular, the following assertions hold*:
  (1) *A is a UFD if and only if X is isomorphic to a hypersurface $x^2 + y^3 + z^5 = 0$ in $\mathbf{A}^3$;*
  (2) *X is affine-ruled if and only if G is cyclic.*

Concerning the topology of an open surface $T = W - \mathrm{Sing}(W)$ with a log del Pezzo surface W, there are several results due to Miyanishi and Zhang [54, 55], Zhang

[75], Gurjar and Zhang [27, 28] and Fujiki, Kobayashi, and Lu [22]. In particular, the following result of Gurjar and Zhang is noteworthy.

THEOREM 9. *Let W be a log del Pezzo surface and let $T = W - \mathrm{Sing}(W)$. Then $\pi_1(T)$ is a finite group.*

Fujita [18] made a quite extensive observation on the topology of open algebraic surfaces, where the crux is the case $\kappa(X) = 0$. He classified affine surfaces X with $\kappa(X) = 0$ and $\mathrm{Pic}(X) \otimes_{\mathbf{Z}} \mathbf{Q} = (0)$ under some minimality condition on the boundary divisor, though we cannot reproduce his results here for lack of space.

As for the case $\kappa(X) = 1$, the following result of Kawamata [33] is decisive.

THEOREM 10. *Let X be an open algebraic surface of $\kappa(X) = 1$. Then X has either an elliptic fibration or an $\mathbf{A}_*^1$-fibration.*

The structure of an open algebraic surface with an $\mathbf{A}_*^1$-fibration is parallel to that of a complete algebraic surface with an elliptic fibration. (See Miyanishi [43] and Fujita [18].)

In the case where $\kappa(X) = 2$, X is called an algebraic surface of *general type*. For example, the Ramanujam surface and other contractible surfaces treated in §2 are surfaces of general type. There are important results due to Kawamata [33] in this case. (See also [43].)

THEOREM 11. *Let X be an open algebraic surface of general type and let (V, D) be a normal completion of X. Set $R := \bigoplus_{n \geq 0} H^0(V, \mathcal{O}(D + K_V)^{\otimes n})$. Then we have*
  (1) *R is a finitely generated, graded $\mathbf{C}$-algebra of dimension 3.*
  (2) *Set $\overline{V}_c := \mathrm{Proj}(R)$. Then $\overline{V}_c$ is a projective normal surface with log canonical singularities. Namely, the singularity on $\overline{V}_c$ is a quotient singular point, an elliptic singular point, a quasi-elliptic singular point (i.e., a quotient of an elliptic singular point under a finite group action), a cuspidal singular point, or a quasi-cuspidal singular point (i.e., the quotient of a cuspidal singular point under a finite group action).*
  (3) *$\overline{V}_c$ is obtained as the image of a rational mapping Φ_n associated with $H^0(V, \mathcal{O}(D + K_V)^{\otimes n})$ for $n \gg 0$.*

The following result of Kobayashi [36] provides a powerful analytic tool for the study of open algebraic surfaces of general type. (See also Kobayashi [34, 35], Kobayashi, Nakamura, and Sakai [81], Miyaoka [41], and Miyanishi and Tsunoda [53].)

THEOREM 12. *With the hypotheses in Theorem 11, the following assertions hold.*
  (1) *Set $\overline{V}_0 = \overline{V}_c - \overline{\Delta}_c - \mathrm{LCS}(\overline{V}_c)$, where $\overline{\Delta}_c$ is the union of divisor components of $\Phi_n(D)$ and $\mathrm{LCS}(\overline{V}_c)$ is the set of all singular points of $\overline{V}_c$ except for quotient singular points. Then $\overline{V}_0$ has a complete, Ricci-negative, Kähler orbifold metric with finite volume, which is unique up to multiplication by positive numbers.*
  (2) *We have an inequality*
$$0 < (K_{\overline{V}_c} + \overline{\Delta}_c)^2 \leq 3 \left\{ e(\overline{V}_0) + \sum_P \left(\frac{1}{|\Gamma(P)|} - 1 \right) \right\},$$
where $K_{\overline{V}_c}$ is the canonical (Weil) divisor of $\overline{V}_c$, $e(\overline{V}_0)$ is the Euler number of $\overline{V}_0$, and $|\Gamma(P)|$ is the order of the local fundamental group of $\overline{V}_0$ with P ranging over all singular points of $\overline{V}_0$.

For a given open algebraic surface X let (V, D) be a normal completion of X. By contracting all log exceptional curves of the first kind it is possible to construct a relatively minimal model of X. This process may introduce log terminal singularities (same as quotient singularities). For an open algebraic surface of general type, by contracting all log exceptional curves of the second kind we may construct a minimal model; this model can be identified with $(\overline{V}_0, \overline{\Delta}_c)$ and may admit log canonical singularities. See Sakai [63] and Gurjar and Miyanishi [24].

Concerning the cancellation problem, we add some remarks. By Fujita and Iitaka [21], the hypothesis $X \times \mathbf{C} \cong Y \times \mathbf{C}$ for nonsingular algebraic surfaces X and Y with nonnegative Kodaira dimensions implies that X is isomorphic to Y. However, in the case of Kodaira dimension $-\infty$, there is the following counterexample of Danielewski [3].

Let $X(n)$ be an affine hypersurface in $\mathbf{A}^3$ defined by an equation $x^n y + z^2 = 1$, where $n \geq 1$. Then we have
(1) $X(n) \times \mathbf{A}^1 \cong X(m) \times \mathbf{A}^1$ for arbitrary $n, m \geq 1$.
(2) (Fieseler) $\pi_1^\infty(X(n)) \cong (\mathbf{Z}/n\mathbf{Z})^2$ if $n \geq 2$, and $\pi_1^\infty(X(n)) \cong \mathbf{Z}/2\mathbf{Z}$ if $n = 1$; see §2 for the fundamental group at infinity.

In fact, $X(1)$ is isomorphic to $\mathbf{P}^1 \times \mathbf{P}^1 - $(diagonal), and $X(n)$ is obtained from $X(1)$ by the nth cyclic covering $x \mapsto x^n$. The surface $X(n)$ has an $\mathbf{A}^1$-fibration induced by $(x, y, z) \mapsto x$. Hence $\kappa(X(n)) = -\infty$, but rank Pic $(X(n)) \otimes \mathbf{Q} = 1$. See tom Dieck [5] and Kraft [38].

§4. Homology planes and related topics

Let X be a $\mathbf{Q}$-homology plane and let (V, D) be a normal completion of X. Then we know the following (cf. [48]):
(1) X is an affine surface;
(2) the geometric genus $p_g(V) = 0$ and the irregularity $q(V) = 0$;
(3) D is simply connected, i.e., every irreducible component of D is isomorphic to $\mathbf{P}^1$ and the dual graph of D is a tree;
(4) Pic(X) is a finite group and $\Gamma(\mathcal{O}_X)^* = \mathbf{C}^*$;
(5) if X is rational, then $H_i(X; \mathbf{Z}) = (0)$ for $i \geq 2$ and Pic$(X) \cong H_1(X; \mathbf{Z}) \cong H^2(X; \mathbf{Z}) \cong $ Coker$(H^2(V; \mathbf{Z}) \to H^2(D; \mathbf{Z}))$.

If $\kappa(X) = -\infty$, then X has an $\mathbf{A}^1$-fibration $\rho: X \to \mathbf{A}^1$ whose fibers are all isomorphic to $\mathbf{A}^1$ counted with some multiplicities. If $\{f^{-1}(P_i): 1 \leq i \leq m\}$ exhausts all multiple fibers and μ_i is the multiplicity of $f^{-1}(P_i)$, then $H_1(X; \mathbf{Z}) = \bigoplus_{i=1}^m \mathbf{Z}/\mu_i \mathbf{Z}$. Hence, ρ has no multiple fibers provided $H_1(X; \mathbf{Z}) = (0)$. This entails that a homology plane of $\kappa(X) = -\infty$ is isomorphic to the affine plane $\mathbf{A}^2$. See [48] for the above result. tom Dieck [5] studied Danielewski's counterexample to the cancellation problem and found $\mathbf{Q}$-homology planes of Kodaira dimension $-\infty$ that are also counterexamples to the cancellation problem.

In the case where $\kappa(X) = 0$, we can make use of Fujita's classification mentioned in §3. According to his notation, X is $H[k, -k]$ ($k \geq 1$), $Y\{3, 3, 3\}$, $Y\{2, 4, 4\}$, or $Y\{2, 3, 6\}$. Hence $H_1(X; \mathbf{Z})$ is accordingly $\mathbf{Z}/4k\mathbf{Z}, \mathbf{Z}/9\mathbf{Z}, \mathbf{Z}/8\mathbf{Z}$, or $\mathbf{Z}/6\mathbf{Z}$. This entails that there are no homology planes of $\kappa(X) = 0$. See [48].

In the case where $\kappa(X) = 1$, there exists an $\mathbf{A}_*^1$-fibration $\rho: X \to C$ with C isomorphic to $\mathbf{P}^1$ or $\mathbf{A}^1$. If X is a homology plane, then $C \cong \mathbf{P}^1$ and the fibration ρ is untwisted. The construction and structure of X are described completely by embedding X (together with the fibration ρ) into a $\mathbf{P}^1$-fibration $p: V \to C$ and by indicating the blow-up process for obtaining V from a relatively minimal ruled surface. In particular, all homology planes of Kodaira dimension 1 as well as contractible surfaces of Kodaira dimension 1 are classified. See [23]. tom Dieck and Petrie [10] showed that every contractible surface of Kodaira dimension 1 is obtained from a *basic* contractible surface by a *standard k-fold blow-up process* applied to a point on the boundary component at infinity, where basic contractible surfaces are isomorphic to affine hypersurfaces $V(n,a)$ in $\mathbf{A}^3$ defined by the equations

$$f_{n,a} = \frac{1}{z}\{(xz+1)^n - (yz+1)^a\}$$

with coprime integers n and a satisfying $n > a > 1$.

Furthermore, Petrie [58] observed that a contractible homology plane of Kodaira dimension 0 or 1 has no nontrivial automorphisms, and asked whether or not this should also hold for homology planes of general type.

If X is a **Q**-homology plane with a group G of automorphisms, the quotient surface X/G has vanishing reduced **Q**-homologies, though X/G may acquire quotient singularities. This motivates one to introduce log **Q**-homology planes and log homology planes.

A normal algebraic surface Y is called a *log **Q**-homology plane* (resp. a *log homology plane*) if Y has only quotient singularities and $H_i(Y;\mathbf{Q}) = (0)$ (resp. $H_i(Y;\mathbf{Z}) = (0)$) for all $i > 0$. The structure of logarithmic **Q**-homology planes with $\mathbf{A}_*^1$-fibrations is given in [48]. It is noteworthy that if Y is a log homology plane with an $\mathbf{A}_*^1$-fibration $\rho: Y \to C$ and $\kappa(Y) \geq 0$, then $\kappa(Y) = 1$, $C \cong \mathbf{P}^1$, the fibration ρ is untwisted, and Y has at most one singular point that is necessarily a cyclic quotient singularity.

In the case where $\kappa(X) = 2$, there is no complete classification. The above observation of Petrie motivated a search for homology planes with nontrivial automorphisms other than $\mathbf{A}^2$. Note that the automorphism group of a log **Q**-homology plane of general type is a finite group. Miyanishi and Sugie [49] considered **Q**-homology planes with $\mathbf{C}^{**}$-fibrations, where $\mathbf{C}^{**}$ is the affine line minus two points and a $\mathbf{C}^{**}$-fibration is a substitute in the non-complete case for a genus 2 fibration in the complete case and found infinitely many homology planes with nontrivial automorphisms after classifying **Q**-homology planes with $\mathbf{C}^{**}$-fibrations.[2] (See also Zaidenberg [72].)

tom Dieck [8] took a different approach for constructing homology planes of general type. Since such a surface X is rational by Gurjar and Shastri [26], we can find a normal completion (V,D) of X so that V has a birational morphism $f: V \to \mathbf{P}^2$. X is said to result from a *line arrangement* on $\mathbf{P}^2$ if the direct image $f_*(D)$ is a union of lines. tom Dieck [8] classified all line arrangements on $\mathbf{P}^2$ that yield (**Q**-) homology planes under mild restrictions on line arrangements. See also tom Dieck and Petrie [13]. In [7], tom Dieck also considered an arrangement of plane curves enjoying special intersection properties, e.g., the Steiner quartic and its bi-tangent, for constructing a homology plane with $\mathbf{Z}/3\mathbf{Z}$-symmetry. See also [11]. In [12], tom Dieck and Petrie extended the arguments to arrangements of curves on relatively minimal rational surfaces.

---

[2] (Added to the English translation) Sugie and the author recently showed that if X is a homology plane of general type with a $\mathbf{C}^{**}$-fibration then $c_1(X)^2 < 2c_2(X)$, where $c_i(X)$ ($i = 1, 2$) is the Chern number of X. This is the non-complete version of Xiao-Gang's result for a nonsingular projective surface of general type with a genus 2 fibration (see Persson [79]).

Besides various constructions of homology planes of general type, Zaidenberg [70] found that those surfaces have the following remarkable property. An analytic proof due to Gurjar and Miyanishi [24] utilizes the inequality of Miyaoka-Yau type due to Kobayashi, Nakamura, and Sakai as is given in Theorem 12. See also Miyanishi and Tsunoda [53].

THEOREM 13. *Let X be a log $\mathbf{Q}$-homology plane. Then we have*
(1) if $\kappa(X) = 2$, then X contains no contractible curves, i.e., those curves homeomorphic to $\mathbf{A}^1$.
(2) If X is a log homology plane with $\kappa(X) = 1$, then X contains a unique contractible curve, which is the affine line containing at most one (cyclic) quotient singular point of X.

Recently, Gurjar and Parameswaran [77] proved that if X is a $\mathbf{Q}$-homology plane with $\kappa(X) = 0$ then X has at most two contractible curves; in fact, they obtained more precise results.

Flenner and Zaidenberg [16] considered logarithmic (infinitesimal) deformations of normal completions (V, D) of certain $\mathbf{Q}$-homology planes X and computed the dimensions of the spaces $H^i(V, \Theta(D))$.

There are several trials for finding topologically contractible, nonsingular affine n-folds with $n \geq 3$. Let X be a nonsingular affine hypersurface in $\mathbf{A}^{n+1}$. If X is topologically contractible and $n \geq 3$, then X is diffeomorphic to $\mathbf{R}^{2n}$ by the h-cobordism theory (cf. [31]). An example of such hypersurfaces is given by Dimca [15] and is defined as follows:

$$x_1^a x_2^{d-a} + x_2 x_3^{d-1} + \cdots + x_{n-1} x_n^{d-1} + x_n + x_{n+1}^d = 0,$$

where $0 < a < d-1$, $(a, d) = (a, d-1) = 1$, $n \geq 3$, and $n \equiv 1 \pmod 2$. Another example, due to Kaliman [31], is defined by

$$x_1 + x_1^4 x_2 + x_2^2 x_3^3 + x_4^3 + x_5^4 = 0.$$

One can hope that some of these hypersurfaces are not isomorphic to $\mathbf{A}^n$, thus giving an exotic structure on $\mathbf{R}^{2n}$. Even if they are isomorphic to $\mathbf{A}^n$, one may still ask the following interesting question, which is often called the conjecture of Abhyankar-Sathaye:

Let X be an affine hypersurface in $\mathbf{A}^{n+1}$. Suppose that X is defined by $f = 0$ and that X is isomorphic to $\mathbf{A}^n$. Do there exist elements $f_1, \ldots, f_{n+1} \in \mathbf{C}[x_1, \ldots, x_{n+1}]$ such that $\mathbf{C}[x_1, \ldots, x_{n+1}] = \mathbf{C}[f_1, \ldots, f_{n+1}]$ and $f_1 = f$?

If $n = 1$, the answer is affirmative by Abhyankar and Moh [2]; see Miyanishi [46] for a geometric proof and Rudolph [60] for a topological proof.

Finally, we remark that the conjecture of Abhyankar-Sathaye is deeply related to the linearization problem of C^*-actions on $\mathbf{C}^3$. See Kraft [38] for a survey of this problem and Koras and Russell [37] for a recent result.

References

1. A. Andreotti and T. Frankel, *The Lefschetz theorem on hyperplane sections*, Ann. of Math. 69 (1959), 713–717.
2. S. S. Abhyankar and T. T. Moh, *Embeddings of the line in the plane*, J. Reine Angew. Math. 276 (1975), 148–166.
3. W. Danielewski, *On the cancellation problem and automorphism group of affine algebraic varieties*, preprint.

4. P. Deligne, *Le groupe fondamental du complément d'une courbe plane n'ayant que des points doubles ordinaires est abélian*, Séminaire Bourbaki, 32e anné, 1979/80, no. 543, Lecture Notes in Math., vol. 842, Springer-Verlag, Berlin-Heidelberg-New York, 1981, pp. 1–10.
5. T. tom Dieck, *On the topology of complex surfaces with an action of the additive group of complex numbers*, preprint, Mathematica Göttingensis 28 (1992).
6. T. tom Dieck, *Homology planes without cancellation property*, Arch. Math. 59 (1992), 105–114.
7. T. tom Dieck, *Symmetric homology planes*, Math. Ann. 286 (1990), 143–152.
8. T. tom Dieck, *Linear plane divisors of homology planes*, J. Fac. Sci. Univ. Tokyo, Sect. IA Math. 37 (1990), 33–69.
9. T. tom Dieck and T. Petrie, *Homology planes: An announcement and survey*, Topological Methods in Algebraic Transformation Groups, Progress in Math. 80 (1989), 27–48.
10. T. tom Dieck and T. Petrie, *Contractible affine surfaces of Kodaira dimension 1*, Japan J. Math. 16 (1990), 147–169.
11. T. tom Dieck and T. Petrie, *Homology planes and algebraic curves*, I, preprint, Mathematica Göttingensis 6 (1989).
12. T. tom Dieck and T. Petrie, *Homology planes and algebraic curves*, Osaka J. Math. 30 (1993), 855–886.
13. T. tom Dieck and T. Petrie, *Arrangements of lines with tree resolutions*, Arch. Math. 56 (1991), 189–196.
14. T. tom Dieck and T. Petrie, *Optimal rational curves and homology planes*, preprint, Mathematica Göttingensis 9 (1992).
15. A. Dimca, *Hypersurfaces in $\mathbf{C}^{2n}$ diffeomorphic to $\mathbf{R}^{4n-2}$ ($n \geq 2$)*, preprint, Max-Planck Inst. für Math.
16. H. Flenner and M. Zaidenberg, *$\mathbf{Q}$-acyclic surfaces and their deformations*, preprint, Mathematica Göttingensis 23 (1992).
17. T. Fujita, *On Zariski problem*, Proc. Japan Acad., Ser. A, Math. Sci. 55 (1979), 106–110.
18. T. Fujita, *On the topology of non-complete algebraic surfaces*, J. Fac. Sci. Univ. Tokyo, Sect. IA Math. 29 (1982), 503–566.
19. T. Fujita, *Classification of noncomplete algebraic varieties*, Proc. Sympos. Pure Math. 46 (1987), 417–423.
20. W. Fulton and J. Hansen, *A connectedness theorem for projective varieties, with applications to intersections and singularities of mappings*, Ann. of Math. 110 (1979), 159–166.
21. T. Fujita and S. Iitaka, *Cancellation theorem for algebraic varieties*, J. Fac. Sci. Univ. Tokyo, Sect. IA Math. 24 (1977), 123–127.
22. A. Fujiki, R. Kobayashi, and S. Lu, *On the fundamental group of certain normal surfaces*, Saitama Math. J. 11 (1993), 15–20.
23. R. V. Gurjar and M. Miyanishi, *Affine surfaces with $\bar{\kappa} \leq 1$*, Algebraic Geometry and Commutative Algebras, Kinokuniya, Tokyo, 1987, pp. 99–124.
24. R. V. Gurjar and M. Miyanishi, *Affine lines on logarithmic $\mathbf{Q}$-homology planes*, Math. Ann. 294 (1992), 463–482.
25. R. V. Gurjar and A. R. Shastri, *A topological characterization of $\mathbf{C}^2/G$*, J. Math. Kyoto Univ. 25 (1985), 767–773.
26. R. V. Gurjar and A. R. Shastri, *On the rationality of complex homology 2-cells*, I and II, J. Math. Soc. Japan 41 (1989), 37–56 and 175–212.
27. R. V. Gurjar and D.-Q. Zhang, *π_1 of smooth parts of a log del Pezzo surface is finite*, I and II, Max Planck Inst. für Math., preprint.
28. R. V. Gurjar and D.-Q. Zhang, *On the fundamental group of some open rational surfaces*, preprint.
29. S. Iitaka, *On logarithmic Kodaira dimension of algebraic varieties*, Complex Analysis and Algebraic Geometry, Iwanami, Tokyo, 1977, pp. 175–189.
30. T. Kambayashi, *On Fujita's strong cancellation theorem for the affine plane*, J. Fac. Sci. Univ. Tokyo, Sect. IA Math. 23 (1980), 535–548.
31. S. Kaliman, *Smooth contractible hypersurfaces in $\mathbf{C}^n$ and exotic structures on $\mathbf{C}^3$*, preprint.
32. Y. Kawamata, *On deformations of compactifiable complex manifolds*, Math. Ann. 235 (1978), 247–265.
33. Y. Kawamata, *On the classification of non-complete algebraic surfaces*, Proc. Copenhagen Summer Meeting in Algebraic Geometry, Lecture Notes in Math., vol. 732, Springer-Verlag, Berlin-Heidelberg-New York, 1979, pp. 215–232.
34. R. Kobayashi, *Einstein-Kähler metrics on open algebraic surfaces of general type*, Tôhoku Math. J. 37 (1985), 43–77.

35. R. Kobayashi, *Einstein-Kähler V-metrics on open Satake V-surfaces with isolated quotient singularities*, Math. Ann. **272** (1985), 385–398.
36. R. Kobayashi, *Uniformization of complex surfaces*, Adv. Stud. Pure Math. **18** (1990), 313–394.
37. M. Koras and P. Russell, *On linearization of "good" C^*-actions on C^3*, Proc. Can. Math. Soc. Conf. **10** (1989), 93–102.
38. H. P. Kraft, *Algebraic automorphisms of affine space*, Topological Methods in Algebraic Transformation Groups, Progress in Math. **80** (1989), 81–105.
39. J. Milnor, *Morse Theory*, Princeton Univ. Press, Princeton, NJ, 1963.
40. J. Milnor, *Lectures on the h-Cobordism Theorem*, Princeton Univ. Press, Princeton, NJ, 1965.
41. Y. Miyaoka, *The maximal number of quotient singularities on surfaces with given numerical invariants*, Math. Ann. **26** (1984), 159–171.
42. M. Miyanishi, *An algebraic characterization of the affine plane*, J. Math. Kyoto Univ. **15** (1975), 169–184.
43. M. Miyanishi, *Non-complete algebraic surfaces*, Lecture Notes in Math., vol. 857, Springer-Verlag, Berlin-Heidelberg-New York, 1981.
44. M. Miyanishi, *Regular subrings of a polynomial ring*, Osaka J. Math. **17** (1980), 329–338.
45. M. Miyanishi, *Normal affine subalgebras of a polynomial ring*, Algebraic and Topological Theories, Kinokuniya, Tokyo, 1985, pp. 37–51.
46. M. Miyanishi, *Analytic irreducibility of certain curves on a nonsingular affine rational surface*, Proc. Internat. Sympos. on Algebraic Geometry, Kinokuniya, Tokyo, 1977, pp. 575–587.
47. M. Miyanishi and T. Sugie, *Affine surfaces containing cylinderlike open sets*, J. Math. Kyoto Univ. **20** (1980), 11–42.
48. M. Miyanishi and T. Sugie, *Homology planes with quotient singularities*, J. Math. Kyoto Univ. **31** (1991), 755–788.
49. M. Miyanishi and T. Sugie, *Q-homology planes with C^{**}-fibrations*, Osaka J. Math. **28** (1991), 1–26.
50. M. Miyanishi and S. Tsunoda, *Non-complete algebraic surfaces with logarithmic Kodaira dimension $-\infty$ and with non-connected boundaries at infinity*, Japan J. Math. **10** (1984), 195–242.
51. M. Miyanishi and S. Tsunoda, *Logarithmic del Pezzo surfaces of rank one with non-contractible boundaries*, Japan J. Math. **10** (1984), 271–319.
52. M. Miyanishi and S. Tsunoda, *Open algebraic surfaces with Kodaira dimension $-\infty$*, Proc. Sympos. Pure Math. **46** (1987), 435–450.
53. M. Miyanishi and S. Tsunoda, *Absence of the affine lines on the homology planes of general type*, J. Math. Kyoto Univ. **32** (1992), 443–450.
54. M. Miyanishi and D.-Q. Zhang, *Gorenstein log del Pezzo surfaces of rank one*, J. Algebra **118** (1988), 63–84.
55. M. Miyanishi and D.-Q. Zhang, *Gorenstein log del Pezzo surfaces*, II, J. Algebra **156** (1993), 183–193.
56. J. A. Morrow, *Minimal compactifications of C^2*, Rice Univ. Studies **59** (1973), 97–112.
57. R. Palais, *Extending diffeomorphisms*, Proc. Amer. Math. Soc. **11** (1960), 274–277.
58. T. Petrie, *Algebraic automorphisms of smooth affine surfaces*, Invent. Math. **138** (1989), 355–378.
59. C. P. Ramanujam, *A topological characterization of the affine plane as an algebraic variety*, Ann. of Math. **94** (1971), 69–88.
60. L. Rudolph, *Embeddings of the line in the plane*, J. Reine Angew. Math. **337** (1982), 113–118.
61. P. Russell, *On affine-ruled rational surfaces*, Math. Ann. **255** (1981), 287–302.
62. F. Sakai, *Kodaira dimensions of complements of divisors*, Complex Analysis and Algebraic Geometry, Iwanami, Tokyo, 1977, pp. 239–257.
63. F. Sakai, *Classification of normal surfaces*, Proc. Sympos. Pure Math. **46** (1987), 451–465.
64. S. Smale, *Generalized Poincaré conjecture in dimensions greater than 4*, Ann. of Math. **64** (1956), 399–405.
65. T. Sugie, *On a characterization of surfaces containing cylinderlike open sets*, Osaka J. Math. **17** (1980), 339–362.
66. T. Sugie, *On Petrie's problem concerning homology planes*, J. Math. Kyoto Univ. **30** (1990), 317–342.
67. S. Tsunoda, *Structure of open algebraic surfaces*, I, J. Math. Kyoto Univ. **23** (1983), 95–125.
68. S. Tsunoda and M. Miyanishi, *The structure of open algebraic surfaces*, II, Proc. Taniguchi Sympos. on Algebraic Geometry, Progress in Math., vol. 39, Birkhäuser, Boston-Basel-Stuttgart, 1983, pp. 499–544.
69. A. Van de Ven, *Analytic compactifications of complex homology cells*, Math. Ann. **147** (1962), 189–204.
70. M. G. Zaidenberg, *Isotrivial families of curves on affine surfaces, and the characterization of the affine plane*, Izv. Akad. Nauk SSSR Ser. Mat. **51** (1987), 534–567, 688 (Russian); English translation: Math. USSR Izvestija **30** (1988), 503–532.

71. M. G. Zaidenberg, *An analytic cancellation theorem and exotic algebraic structures on $\mathbf{C}^n$, $n \geq 3$*, Max-Planck Inst. für Math., preprint.
72. M. G. Zaidenberg, *On Ramanujam surfaces, $\mathbf{C}^{**}$-families and exotic algebraic structures on $\mathbf{C}^n$, $n \geq 3$*, Soviet Math. Dokl. **42** (1991), 636–640.
73. O. Zariski, *On the problem of existence of algebraic functions of two variables possessing a given branch curve*, Amer. J. Math. **51** (1929), 305–328.
74. D.-Q. Zhang, *Logarithmic del Pezzo surfaces of rank one with contractible boundaries*, Osaka J. Math. **25** (1988), 461–497.
75. D.-Q. Zhang, *Logarithmic del Pezzo surfaces with rational double and triple points*, Tôhoku Math. J. **41** (1989), 399–452.
76. D.-Q. Zhang, *Logarithmic Enriques surfaces*, J. Math. Kyoto Univ. **31** (1991), 419–466.
77. R. V. Gurjar and A. J. Parameswaran, *Affine lines on $\mathbf{Q}$-homology planes*, Tata Institute of Fundamental Research, preprint.
78. R. V. Gurjar and A. J. Parameswaran, *On the Euler characteristic of open surfaces*, Tata Institute of Fundamental Research, preprint.
79. U. Persson, *An introduction to the geography of surfaces of general type*, Proc. Sympos. Pure Math. **46** (1987), 195–218.
80. K.-H. Fieseler, *On complex affine surfaces with $\mathbb{C}^+$-action*, Comment. Math. Helvetici **69** 1994, 5–27.
81. R. Kobayashi, S. Nakamura, and F. Sakai, *A numerical characterization of ball quotients for normal surfaces with branch loci*, Proc. Japan Acad. **65**, Ser. A (1989), 238–241.

DEPARTMENT OF MATHEMATICS, FACULTY OF SCIENCE, OSAKA UNIVERSITY, TOYONAKA, OSAKA 560, JAPAN

Translated by M. MIYANISHI

Recent Topics on Toric Varieties

Tadao Oda

§1. Introduction

The theory of toric varieties, or *toric geometry* for short, establishes a close relationship between the geometry of convex bodies with respect to lattices and algebraic geometry. It was discovered at the beginning of the 1970s by Demazure [18], Mumford et al. [38], Satake [60], and Miyake and Oda [42]. The basic dictionary of the theory relating the geometry of convex bodies with respect to lattices to algebraic geometry was already completed at the time of the inception, thanks in part to the timely discovery by Sumihiro [63] of a key result on the algebraic actions of linear algebraic groups. It should be noted that what we now call toric varieties were called torus embeddings at that time.

As we briefly recall in §3, a quite simple-minded basic principle enables us to relate the geometry of convex bodies with respect to lattices to algebraic geometry. Nevertheless, toric geometry keeps on growing rapidly even more than twenty years after its inception, with amazingly many new discoveries being made almost every day. As a testimony to this, the new number 14M25 for "Toric varieties, Newton polyhedra" was added at the time of the 1991 revision of the *Mathematics Subject Classification* scheme, which is used by the *Mathematical Reviews* and *Zentralblatt für Mathematik* in classifying mathematical papers. Consequently, we no longer have the earlier difficulty in classifying or searching for papers on toric varieties, which were scattered throughout because of the diversity of topics related to toric varieties.

Due to the limitation on the length of this article, we here try to give a brief but comprehensive survey of inter-related recent topics, leaving more detailed explanation to previously published articles and books such as Ash, Mumford, Rapoport, and Tai [1], Danilov [15], Ewald [22], [23], Fulton [25], Kempf, Knudsen, Mumford, and Saint-Donat [38], and Oda [47], [48], [49], [50], [52]. We also limit the references to the barest minimum and refer the reader for more information to the above literature as well as to those articles and books cited therein. ([54] might be helpful for those who try to get to know toric varieties for the first time.) At the time of the writing of the Japanese version of this article, many of the cited papers were not yet published. Their preprints are available electronically from ALGEBRAIC

1991 *Mathematics Subject Classification*. Primary 14M25; Secondary 14F40, 14F32, 32S60.

Originally submitted on March 31, 1994 and published in Sūgaku, Math. Soc. Japan Vol. 46 (1994), 35–47.

Partly supported by the Grants-in-Aid for Cooperative Research as well as Scientific Research, the Ministry of Education, Science and Culture, Japan.

©1996, American Mathematical Society

GEOMETRY E-PRINTS. For details, send a blank e-mail with "Subject: help" to alg-geom@publications.math.duke.edu.

§2. Multifaceted advances

The multifaceted advances related to toric geometry could be classified as follows:
(1) advances and generalizations of toric geometry itself;
(2) applications;
(3) from the geometry of convex bodies with respect to lattices to algebraic geometry;
(4) from algebraic geometry to the geometry of convex bodies with respect to lattices.

Here are some examples in each category, although the division into these four categories is not so clear-cut. It goes without saying that there are many other important examples.

2.1. Advances and generalizations of toric geometry itself. Toric geometry is based on a very simple principle. Nevertheless, it has room for surprising advances and generalizations as the following examples show.

2.1.1. *Homogeneous coordinate rings.* Cox [13] recently found a surprising result to the effect that homogeneous coordinate rings can be defined for general toric varieties and play roles similar to those in the case of projective varieties. We give more details in §4.3.

2.1.2. *Toroidal embeddings and logarithmic schemes.* Toroidal embeddings are varieties that are locally toric varieties and seem to have been a motivation for Mumford et al. [38], [1] and Satake [60] to introduce toric varieties in the first place. They play important roles in compactifications of moduli spaces and locally symmetric varieties. Danilov [17] and Ishida [31] generalized toroidal embeddings. Kato [37] developed the ideas of Fontaine and Illusie to establish the theory of logarithmic schemes, which parallels and generalizes Grothendieck's theory of schemes and is expected to play a crucial role in arithmetic geometry. Kajiwara [36] used logarithmic schemes to reformulate the results in [56] and subsequent results due to Ishida concerning compactifications of the generalized Jacobian varieties of singular algebraic curves (see also Caporaso [12]). These results are related to tilings of real affine spaces as well as to degenerations of algebraic varieties.

2.2. Applications. Toric varieties turn out to be useful outside algebraic geometry. Here are some examples:

2.2.1. *Generalized hypergeometric functions and holonomic systems of q-difference equations.* Toric varieties play an important role in the rapid recent advances on generalized hypergeometric functions and holonomic systems of q-difference equations. For lack of space here, we refer the reader to a slightly outdated survey [52, §2.4, §2.6]. We only add that Dabrowski [14] finally succeeded in proving the normality of the closure in G/P of maximal torus orbits (cf. [52, p. 430]).

2.2.2. *Einstein-Kähler manifolds.* Fano manifolds are those with ample anticanonical line bundles. It is an interesting unsolved problem in differential geometry to determine which of those manifolds admit Einstein-Kähler metrics. As we explain in §4.4, toric Fano manifolds among them can be dealt with from the viewpoint of convex polytopes as in Mabuchi [40]. Nakagawa [45], [46] determined all those that admit Einstein-Kähler metrics among the four-dimensional toric Fano manifolds classified by Batyrev.

2.2.3. *Toric varieties as ambient spaces.* Affine spaces, projective spaces, and weighted projective spaces, which play basic roles as ambient spaces, are special cases of toric varieties. It is quite likely that more general toric varieties with the homogeneous coordinate rings of Cox [13] mentioned above play increasingly important roles as ambient spaces. As we explain in §4.4 in more detail in connection with the mirror symmetry in mathematical physics, toric Fano varieties are ambient spaces of the Calabi-Yau varieties studied by Roan and Batyrev. We refer the reader to [50, §2.4, p. 98] for a noncompact toric variety used by Ishida as a convenient ambient space.

2.3. From the geometry of convex bodies with respect to lattices to algebraic geometry. Toric varieties provide convenient introductory material for algebraic geometry, since we can use toric varieties to produce interesting examples and phenomena in algebraic geometry from rather elementary results on the geometry of convex bodies with respect to lattices. Sometimes, serious work on the geometry of convex bodies with respect to lattices enables us to solve interesting problems in algebraic geometry as in the following example.

2.3.1. *Factorizations of birational maps.* It is one of the basic unsolved problems in birational geometry to determine whether a given birational transformation between compact nonsingular varieties can be factored into a composite of blow-ups and blow-downs along nonsingular centers. To get a good feeling in the general case, we posed in [47, §9] an analog asking if an equivariant birational transformation between compact nonsingular toric varieties can be factored as a composite of equivariant blow-ups and blow-downs along nonsingular centers. By toric geometry, the problem is reduced to one on star subdivisions of nonsingular fans. As we pointed out in [50, §1.7], a strong factorization is possible in dimension two, while Danilov [16] showed that a weak factorization is possible in dimension three. More recently, Włodarczyk [64] and Morelli [43] independently showed that a weak factorization is always possible in all dimensions, although the present author has not yet checked the details of the proofs.

2.4. From algebraic geometry to the geometry of convex bodies with respect to lattices. Through toric geometry we can very often derive from algebraic geometry interesting results on the geometry of convex bodies with respect to lattices. It is no exaggeration to say that each theory or theorem in algebraic geometry leads in this way to results on convex bodies and lattices. The results could often be quite unexpected, and direct proofs in convex geometry are to be found later on. Here are some examples.

2.4.1. *Hilbert polynomials, Ehrhart polynomials and their equivariant versions.* As we explain in §3.5, a projective toric variety is obtained from an integral convex polytope, that is, a polytope with lattice points as vertices. The Ehrhart polynomial of the integral convex polytope defined in terms of the cardinality of the lattice points in it turns out to coincide with the Hilbert polynomial of the corresponding projective toric variety. The Serre duality for the projective toric variety corresponds to an interesting reciprocity on the lattice points in the integral convex polytope. We refer the reader to Hibi [30] for more details on recent combinatorial studies.

Brion [9] applied to projective toric varieties not just the Riemann-Roch type formula as above but the Lefschetz-Riemann-Roch formula in equivariant K-theory to obtain more interesting results on the lattice points in an integral convex polytope, which Ishida [32] proved directly later and applied to the duality for Tsuchihashi cusp singularities (cf. [33]). More recent developments are explained in Brion [10], [11] and Sardo Infirri [59].

2.4.2. *The Hodge index theorem and the isoperimetric inequalities.* Teissier used toric geometry to prove the Hodge index theorem and the isoperimetric inequalities (see, e.g., [50, §2.4, §A.4]). As we mention in §2.4.3, McMullen [41] generalized the strong Lefschetz theorem in the context of convex geometry and succeeded in obtaining the Hodge-Riemann-Minkowski quadratic inequalities among the mixed volumes as well as a generalization of the Alexandrov-Fenchel inequality.

2.4.3. *The strong Lefschetz theorem.* Ever since Stanley used the strong Lefschetz theorem to prove the so-called g-theorem characterizing the number of faces of simplicial convex polytopes, there have been attempts (cf., e.g., [51]) to prove the strong Lefschetz theorem directly in the context of convex geometry. McMullen [41] seems to have succeeded at last in giving a direct proof of the strong Lefschetz theorem.

As we explain in [51] and §2.4.4, however, the direct proof of the part necessary for the proof of the g-theorem was also given by Ishida [35, the second diagonal theorem and (4.16)] in connection with toric intersection cohomology.

2.4.4. *Toric intersection cohomology.* The intersection cohomology theory due to Goresky and MacPherson [28], [29] and to Beilinson, Bernstein, and Deligne [8] provides an effective means for the study of complex analytic spaces associated with complex singular algebraic varieties such as general toric varieties. Especially powerful is the decomposition theorem for the intersection cohomology.

As was shown by Denef and Loeser [19], Fieseler [24], and Stanley [61], [62], the toric intersection cohomology, that is, the intersection cohomology of complex analytic spaces associated with toric varieties, gives rise, through the decomposition theorem, to new invariants for and insights into *rational* convex polytopes (i.e., convex polytopes with all the vertices having rational coordinates) and fans. It is thus desirable to describe the toric intersection cohomology in a direct algebraic manner in terms of fans, and then prove the decomposition theorem directly. After attempts such as the algebraic de Rham theorem in [53], Ishida [34], [35] finally succeeded. Not only could he give a completely algebraic description of toric intersection cohomology with respect to arbitrary perversity, but he also proved important results including the decomposition theorem with respect to barycentric subdivisions as well as the first and second diagonal theorems. It is worth noting here that rationality in the definition of fans plays a crucial role in this algebraic description of Ishida's, let alone in the direct application of the usual intersection cohomology to toric varieties. To obtain good combinatorial invariants valid for convex polytopes that are not necessarily rational, we have yet to find an analog of Ishida's description for "convex cone decompositions" in [51], which are fans without the rationality assumption.

2.4.5. *The toric Mori theory and its generalization.* As we see in §§4.1 and 4.2, a generalization exists that is common both to what Reid obtained for toric varieties as a precursor to the Mori theory in birational geometry and to what Gel'fand, Kapranov, and Zelevinskii (cf., e.g., [27]) obtained in connection with hypergeometric functions.

Dolgachev and Hu showed that this generalization is closely related, through the homogeneous coordinate ring of Cox [13] (cf. §4.3), to the current forefront of geometric invariant theory (cf. Dolgachev and Hu [21]).

2.4.6. *Toric intersection theory.* The intersection theory that plays a central role in algebraic geometry gives rise to quite interesting results on fans and polytopes when applied to toric varieties. For details, we refer the reader to Fulton [25] and its references, Park [57], [58], as well as Fulton and Sturmfels [26].

§3. Basic facts on toric geometry

In this section we briefly recall basic facts on toric geometry and introduce new notions and notation that we need in subsequent sections to give a comprehensive survey. For simplicity, we deal only with toric varieties defined over the field $\mathbf{C}$ of complex numbers.

The following simple-minded principle enables us to relate convex geometry and algebraic geometry. To a lattice point $m := (i_1, i_2, \ldots, i_r) \in \mathbf{Z}^r =: M$ we associate the Laurent monomial $t^m := t_1^{i_1} t_2^{i_2} \cdots t_r^{i_r}$ in r variables $t_1, t_2, \ldots, t_r$. Hence to the lattice group M, which is a free $\mathbf{Z}$-module of rank r, we associate the Laurent polynomial ring

$$\mathbf{C}\left[t_1, t_2, \ldots, t_r, \frac{1}{t_1}, \frac{1}{t_2}, \ldots, \frac{1}{t_r}\right]$$

in r variables with coefficients in $\mathbf{C}$. It is essential in toric geometry to make a clear distinction between M and its dual $\mathbf{Z}$-module $N := \mathrm{Hom}_{\mathbf{Z}}(M, \mathbf{Z}) = \mathbf{Z}^r$. We define the bilinear duality pairing $\langle \, , \, \rangle : M \times N \to \mathbf{Z}$ by $\langle m, n \rangle := i_1 j_1 + i_2 j_2 + \cdots + i_r j_r$ for $m := (i_1, \ldots, i_r) \in M$ and $n := (j_1, \ldots, j_r) \in N$. We denote their scalar extensions to the field $\mathbf{R}$ of real numbers by $M_R := M \otimes_{\mathbf{Z}} \mathbf{R}$, $N_{\mathbf{R}} := N \otimes_{\mathbf{Z}} \mathbf{R}$ and $\langle \, , \, \rangle : M_{\mathbf{R}} \times N_{\mathbf{R}} \to \mathbf{R}$. This is the basic setup for toric geometry.

3.1. Fans. A subset $\sigma \subset N_{\mathbf{R}}$ is said to be a *rational convex polyhedral cone* if it coincides with the set of nonnegative linear combinations of some $n_1, n_2, \ldots, n_s \in N$, that is,

$$\sigma = \mathbf{R}_{\geq 0} n_1 + \mathbf{R}_{\geq 0} n_2 + \cdots + \mathbf{R}_{\geq 0} n_s$$
$$:= \{a_1 n_1 + a_2 n_2 + \cdots + a_s n_s \mid a_j \in \mathbf{R}, a_j \geq 0, \forall j\}.$$

Here and elsewhere, we use the notation $\mathbf{R}_{\geq 0} := \{a \in \mathbf{R} \mid a \geq 0\}$, $\mathbf{Z}_{\geq 0} := \{a \in \mathbf{Z} \mid a \geq 0\}$, etc., for simplicity. A rational convex polyhedral cone σ is said to be *strongly convex* if $\sigma \cap (-\sigma) = \{O\}$ holds, where O is the origin of $N_{\mathbf{R}}$. This is also equivalent to saying that σ contains no nonzero $\mathbf{R}$-linear subspaces. A subset $\tau \subset \sigma$ is said to be a *face* of σ (denoted $\tau \prec \sigma$) if there exists an $m_0 \in M$ such that

$$\langle m_0, n \rangle \geq 0, \text{ for every } n \in \sigma, \quad \text{and} \quad \tau = \sigma \cap \{m_0\}^{\perp} := \{n \in \sigma \mid \langle m_0, n \rangle = 0\}.$$

A *fan* for $N = \mathbf{Z}^r$ is a collection Δ of strongly convex rational polyhedral cones in $N_{\mathbf{R}}$ satisfying the following conditions:
(i) if τ is a face of $\sigma \in \Delta$, then $\tau \in \Delta$;
(ii) if $\sigma, \sigma' \in \Delta$, then $\sigma \cap \sigma'$ is a face of both σ and σ'.
$|\Delta| := \bigcup_{\sigma \in \Delta} \sigma$ is called the *support* of Δ. A fan Δ for N is said to be *complete* if $|\Delta| = N_{\mathbf{R}}$.

REMARK. In connection with the toric Mori theory (cf. §4.1), we need to follow Reid in generalizing the definition of a fan. Namely, Δ is a *possibly degenerate* fan if we do not require the strong convexity of each $\sigma \in \Delta$, but only their rationality and (i), (ii). In this case, a $\mathbf{Z}$-submodule $N^0 \subset N$ is easily shown to exist such that N/N^0 is a free $\mathbf{Z}$-module and such that $N_{\mathbf{R}}^0$ is a face of each $\sigma \in \Delta$. Δ is then the pull-back to N of a fan for N/N^0.

3.2. Toric varieties. Toric geometry associates to each fan Δ for $N = \mathbf{Z}^r$ an r-dimensional algebraic variety X, for instance, over the field $\mathbf{C}$ of complex numbers.

The toric variety X is a normal variety endowed with an algebraic action of the r-dimensional algebraic torus $T := N \otimes_{\mathbf{Z}} \mathbf{C}^\times = (\mathbf{C}^\times)^r$, where $\mathbf{C}^\times$ is the multiplicative group of nonzero complex numbers. Moreover, X contains T as an open subset, and the restriction to T of the action of T on X coincides with the group multiplication. Besides, X is known to have exactly one T-orbit $\mathrm{orb}(\sigma) \cong (\mathbf{C}^\times)^{r-\dim \sigma}$ for each $\sigma \in \Delta$.

A *map* from a fan Δ' for $N' = \mathbf{Z}^{r'}$ to a fan Δ for $N = \mathbf{Z}^r$ is a $\mathbf{Z}$-linear map $\varphi \colon N' \to N$ such that its scalar extension $\varphi \colon N'_{\mathbf{R}} \to N_{\mathbf{R}}$ has the property that for each $\sigma' \in \Delta'$ there exists $\sigma \in \Delta$ satisfying $\varphi(\sigma') \subset \sigma$. There then correspond a homomorphism $\varphi_* \colon T' := N' \otimes_{\mathbf{Z}} \mathbf{C}^\times \to T := N \otimes_{\mathbf{Z}} \mathbf{C}^\times$ of algebraic tori and an *equivariant holomorphic map* $\varphi_* \colon X' \to X$ between the corresponding toric varieties.

The fundamental theorem of toric geometry guarantees that these correspondences provide a categorical equivalence ("one-to-one" and "onto"). Moreover, we have a basic dictionary that translates algebro-geometric properties of X faithfully into properties of Δ. Here are some examples: X is *compact* if and only if the corresponding fan Δ is finite and complete. X is *nonsingular* if and only if each $\sigma \in \Delta$ is spanned by a part of a $\mathbf{Z}$-basis of N. More generally, X is an *orbifold* if and only if each $\sigma \in \Delta$ is *simplicial*, that is, σ can be spanned by $\mathbf{R}$-linearly independent elements of N. On the other hand, the equivariant holomorphic map $\varphi_* \colon X' \to X$ is *proper birational* if and only if $\varphi \colon N' \to N$ is an isomorphism, and Δ' is a subdivision of Δ under the identification $N'_{\mathbf{R}} = N_{\mathbf{R}}$ through the isomorphism.

3.3. Linear Gale transforms and support functions. From now on we fix $N = \mathbf{Z}^r$ and denote by $M := \mathrm{Hom}_{\mathbf{Z}}(N, \mathbf{Z})$ its dual $\mathbf{Z}$-module. For simplicity, we deal only with a fan Δ for N that is

(∗) finite with $|\Delta| := \bigcup_{\sigma \in \Delta} \sigma$ being an r-dimensional convex polyhedral cone.

Note, however, that $|\Delta|$ does not have to be strongly convex, so that a complete Δ with $|\Delta| = N_{\mathbf{R}}$ is allowed.

For a fan Δ for N we denote by $\Delta(1) := \{\rho \in \Delta \mid \dim \rho = 1\}$ the set of one-dimensional cones in Δ. For each $\rho \in \Delta(1)$ there exists a unique primitive element $n(\rho) \in N$ such that $\rho = \mathbf{R}_{\geq 0} n(\rho)$. By our assumption (∗) we see that $\{n(\rho) \mid \rho \in \Delta(1)\}$ generates in N a $\mathbf{Z}$-submodule of finite index. Let us now introduce a free $\mathbf{Z}$-module $\widetilde{N}$ having the set of symbols $\{\tilde{n}(\rho) \mid \rho \in \Delta(1)\}$ as a basis, and define a $\mathbf{Z}$-linear map with finite cokernel by

$$\pi \colon \widetilde{N} \to N, \quad \pi(\tilde{n}(\rho)) := n(\rho), \forall \rho \in \Delta(1).$$

Its dual is the injective $\mathbf{Z}$-linear map

$$\pi^* \colon M \to \widetilde{M} := \mathrm{Hom}_{\mathbf{Z}}(\widetilde{N}, \mathbf{Z}), \quad \pi^*(m) := \sum_{\rho \in \Delta(1)} \langle m, n(\rho) \rangle \tilde{m}(\rho),$$

where $\{\tilde{m}(\rho) \mid \rho \in \Delta(1)\}$ is the dual $\mathbf{Z}$-basis of $\widetilde{M}$. The finitely-generated $\mathbf{Z}$-module $\mathscr{M} := \mathrm{coker}(\pi^*) = \widetilde{M}/M$ need not be $\mathbf{Z}$-free. For each $\rho \in \Delta(1)$ we denote by $\mu(\rho)$

the image of $\tilde{m}(\rho)$ in $\mathcal{M}$. On the other hand, $\mathcal{N} := \ker(\pi) = \operatorname{Hom}_{\mathbf{Z}}(\mathcal{M}, \mathbf{Z})$ is a free **Z**-module of finite rank, and we have exact sequences

$$0 \to M \xrightarrow{\pi^*} \widetilde{M} \to \mathcal{M} \to 0, \qquad 0 \leftarrow \operatorname{coker}(\pi) \leftarrow N \xleftarrow{\pi} \widetilde{N} \leftarrow \mathcal{N} \leftarrow 0.$$

$(\mathcal{M}, \{\mu(\rho) \mid \rho \in \Delta(1)\})$ is sometimes called

the *linear Gale transform* of $(N, \{n(\rho) \mid \rho \in \Delta(1)\})$.

We can characterize the transform by the universality of the identity

$$\sum_{\rho \in \Delta(1)} n(\rho) \otimes \mu(\rho) = 0 \quad \text{in } N \otimes_{\mathbf{Z}} \mathcal{M}.$$

A *support function* for Δ is a real-valued function $h \colon |\Delta| \to \mathbf{R}$ that is **Z**-valued on $N \cap |\Delta|$, is positively homogeneous (i.e., $h(\lambda n) = \lambda h(n)$ holds for all $n \in |\Delta|$ and positive λ) and is linear on each $\sigma \in \Delta$. We denote by $\operatorname{SF}(\Delta)$ the **Z**-module consisting of the support functions for Δ. Each $m \in M$ gives rise to a support function linear on the whole of Δ. Consequently, we may regard M as a **Z**-submodule of $\operatorname{SF}(\Delta)$. On the other hand, the **Z**-linear map that sends $h \in \operatorname{SF}(\Delta)$ to $\sum_{\rho \in \Delta(1)} h(n(\rho))\tilde{m}(\rho) \in \widetilde{M}$ is injective by our assumption $(*)$. From now on, we regard $\operatorname{SF}(\Delta)$ as a **Z**-submodule of $\widetilde{M}$ by this map. $\operatorname{SF}(\Delta) \subset \widetilde{M}$ is of finite index if Δ is simplicial.

3.4. Divisors. Let X be the r-dimensional toric variety corresponding to Δ. For each $\rho \in \Delta(1)$, the closure $V(\rho)$ of the T-orbit $\operatorname{orb}(\rho) \cong (\mathbf{C}^\times)^{r-1}$ is an irreducible Weil divisor on X and is invariant under the action of T. The map that sends $\tilde{m}(\rho) \in \widetilde{M}$ to $-V(\rho)$ for each $\rho \in \Delta(1)$ is an isomorphism from $\widetilde{M}$ to the **Z**-module of T-invariant Weil divisors on X. (We use the rather unnatural minus sign in conformity with algebro-geometric convention.) The divisor $D_h := -\sum_{\rho \in \Delta(1)} h(n(\rho))V(\rho)$ corresponding to a support function $h \in \operatorname{SF}(\Delta) \subset \widetilde{M}$ turns out to be a Cartier divisor, and $\operatorname{SF}(\Delta)$ is isomorphic to the **Z**-module of T-invariant Cartier divisors on X. Moreover, the Cartier divisor D_h gives rise to an invertible sheaf $\mathcal{O}_X(D_h)$ on X.

Sending $\mu(\rho) \in \mathcal{M}$ to the linear equivalence class $-v(\rho) := -[V(\rho)]$ of $-V(\rho)$ for each $\rho \in \Delta(1)$, we have an isomorphism from $\mathcal{M}$ to the group of linear equivalence classes of Weil divisors on X, while $\operatorname{SF}(\Delta)/M$ is isomorphic to the Picard group $\operatorname{Pic}(X)$ consisting of the isomorphism classes of invertible sheaves on X.

The invertible sheaf $\mathcal{O}_X(D_h)$ corresponding to $h \in \operatorname{SF}(\Delta)$ is *generated by its sections* if and only if h is *upper convex*, that is,

$$h(n) + h(n') \leq h(n + n'), \quad \text{for all } n, n' \in |\Delta|.$$

This turns out to be also equivalent to D_h being *numerically effective* (nef, for brevity). D_h is *ample* if and only if h is *strictly upper convex with respect to* Δ, that is, for each $\sigma \in \Delta$ there exists an $l_\sigma \in M$ such that

$$h(n) \leq \langle l_\sigma, n \rangle, \text{ for every } n \in |\Delta|, \quad \text{and} \quad \text{"the equality} \Leftrightarrow n \in \sigma\text{"}.$$

As we see in §3.5, support functions and convexity traditionally play important roles in the geometry of convex bodies. Surprisingly, they turn out to correspond to Cartier divisors and ampleness, which play equally important roles in algebraic geometry as well. This coincidence is one of the illustrations of the mysterious harmony existing between both geometries.

3.5. Integral convex polytopes and projective toric varieties.
Here is an important way of systematically obtaining finite complete fans: An r-dimensional convex polytope $P \subset M_{\mathbf{R}}$ is said to be *integral* if all of its vertices belong to the lattice M. We define the *support function* $h_P \colon N_{\mathbf{R}} \to \mathbf{R}$ of P by

$$h_P(n) := \inf_{m \in P} \langle m, n \rangle, \quad n \in N_{\mathbf{R}},$$

and have

$$P = \{m \in M_{\mathbf{R}} \mid \langle m, n \rangle \geq h_P(n), \text{ for every } n \in N_{\mathbf{R}}\}.$$

(Note that we use "inf" in toric geometry in conformity with algebraic geometry, although it is customary in the geometry of convex bodies to use "sup" in the definition of support functions due to Minkowski.) There then exists a unique finite complete fan Δ_P for N such that $h_P \in \mathrm{SF}(\Delta_P)$ and such that h_P is strictly upper convex with respect to Δ_p. The toric variety $\mathbf{P}$ corresponding to this Δ_P is compact, and the Cartier divisor $\mathbf{D}$ corresponding to h_P is ample. Consequently, $\mathbf{P}$ is a projective variety. The pairs $(\mathbf{P}, \mathbf{D})$ thus obtained can be thought of as generalizations of projective spaces and weighted projective spaces, and provide good candidates for ambient spaces as we see in §4.4.

§4. Recent advances in toric geometry

In §2 we gave an overview of recent advances in toric geometry. In this section we give more detailed accounts of some of the salient advances.

4.1. The toric Mori theory.
Suppose that a fan Δ for $N = \mathbf{Z}^r$ satisfies the assumption $(*)$ in §3.3, and define a convex polyhedral cone in $\mathrm{SF}(\Delta)_{\mathbf{R}}$ by

$$\mathrm{UCL}(\Delta) := \{h \colon |\Delta| \to \mathbf{R} \mid h \text{ is positively homogeneous,}$$
$$\text{upper convex, and linear on every } \sigma \in \Delta\}.$$

$h \in \mathrm{SF}(\Delta)$ is strictly upper convex with respect to Δ in the sense of §3.4 if and only if h is in the interior of the convex polyhedral cone $\mathrm{UCL}(\Delta)$. Δ is said to be *quasi-projective* if such an h exists, that is, the dimension of $\mathrm{UCL}(\Delta)$ coincides with that of $\mathrm{SF}(\Delta)_{\mathbf{R}}$.

If that is the case, a face of the convex polyhedral cone $\mathrm{UCL}(\Delta)$ turns out to be necessarily of the form $\mathrm{UCL}(\Delta')$, where Δ' is a (possibly degenerate) fan (cf. the Remark in §3.1) for N such that Δ is a subdivision of Δ'. Correspondingly, we get a *quasi-projective contraction* $X \to X'$. For the proof, we describe the convex polyhedral cone dual to $\mathrm{UCL}(\Delta)$ by means of the geometry of walls (that is, $(r-1)$-dimensional cones) in Δ generalizing Reid's toric precursor of the Mori theory. Furthermore, when Δ is a simplicial finite complete fan, we can relate the convex polyhedral cone dual to $\mathrm{UCL}(\Delta)$ with the Mori cone of curves on the corresponding toric variety. For details, we refer the reader to [55] and [58], where "convex" should be read as "upper convex" in the present context.

4.2. Secondary fans and birational geometry.
Let us fix a quasi-projective fan Δ for $N = \mathbf{Z}^r$ that satisfies the assumption $(*)$ in §3.3. For simplicity, assume further that Δ is *simplicial*. As we saw in §3.3, the index of $\mathrm{SF}(\Delta) \subset \widetilde{M}$ is finite so that we may identify $\mathrm{SF}(\Delta)_{\mathbf{R}} = \widetilde{M}_{\mathbf{R}}$.

A new structure arises when we consider all the quasi-projective fans Δ' for N satisfying $\Delta'(1) \subset \Delta(1)$. Algebro-geometrically, this means that we give beforehand a quasi-projective partial compactification X of an algebraic torus T, and consider all the quasi-projective partial compactifications X' whose codimension-one orbits are allowed to be only among those appearing in the boundary of X.

More precisely, a (possibly degenerate) fan Δ' for N is said to be $\Delta(1)$-*admissible* if Δ' is quasi-projective and satifies $|\Delta'| = |\Delta|$ and $\Delta'(1) \subset \Delta(1)$. We then define a convex polyhedral cone in $\widetilde{M_\mathbf{R}}$ by

$$\mathrm{UCL}^\sim(\Delta') := \left\{ \sum_{\rho \in \Delta(1)} y_\rho \tilde{m}(\rho) \in \widetilde{M_\mathbf{R}} \;\middle|\; \begin{array}{l} \text{There exists some } h' \in \mathrm{UCL}(\Delta') \text{ such that} \\ y_\rho \leq h'(n(\rho)), \text{ for every } \rho \in \Delta(1), \\ y_{\rho'} = h'(n(\rho')), \text{ for every } \rho' \in \Delta'(1) \end{array} \right\},$$

which clearly contains the $\mathbf{R}$-subspace $M_\mathbf{R}$. We define the *Gel'fand-Kapranov-Zelevinskii cone* (the GKZ-cone, for short) for Δ' to be the convex polyhedral cone

$$\mathrm{ucl}(\Delta') := \mathrm{UCL}^\sim(\Delta')/M_\mathbf{R}$$

in the real vector space $\mathscr{M}_\mathbf{R} = \widetilde{M_\mathbf{R}}/M_\mathbf{R} = \sum_{\rho \in \Delta(1)} \mathbf{R}\mu(\rho)$. This GKZ-cone is strongly convex and rational with respect to the $\mathbf{Z}$-module $\mathscr{M}/\mathscr{M}_{\mathrm{tor}}$, where $\mathscr{M}_{\mathrm{tor}}$ denotes the torsion part of $\mathscr{M}$. Not only can we explicitly describe its faces as in §4.1, but

$$\mathrm{GKZ}(N, \Delta(1)) := \{\text{Faces of } \mathrm{ucl}(\Delta') \mid \Delta(1)\text{-admissible}$$
$$\text{(possibly degenerate) fans } \Delta' \text{ for } N\}$$

turns out to be a fan for the $\mathbf{Z}$-module $\mathscr{M}/\mathscr{M}_{\mathrm{tor}}$, which we call the *Gel'fand-Kapranov-Zelevinskii fan* (the GKZ-fan, for short) or the *secondary fan* for $(N, \Delta(1))$. Its support is

$$|\mathrm{GKZ}(N, \Delta(1))| = \mathscr{M}_\mathbf{R}^- := -\sum_{\rho \in \Delta(1)} \mathbf{R}_{\geq 0} \mu(\rho).$$

The maximal dimensional GKZ-fans turn out to be exactly the $\mathrm{ucl}(\Delta')$ for $\Delta(1)$-admissible *simplicial* fans Δ'. Furthermore, when $\Delta(1)$-admissible simplicial fans Δ' and Δ'' are given, $\mathrm{ucl}(\Delta')$ and $\mathrm{ucl}(\Delta'')$ are adjacent to each other through a facet (codimension-one face) if and only if one of them is a *star subdivision* or a *flop* of the other. Thus a $\Delta(1)$-admissible simplicial fan Δ' is obtained from the originally given Δ by a finite succession of taking a star subdivision, an operation inverse to a star subdivision, or a flop. On the one hand, $\mathscr{M}_\mathbf{R}^-$ is strongly convex, when Δ is complete. We thus get a generalization and refinement of Reid's toric Mori theory. As Batyrev pointed out, the result is closely related to the Zariski decomposition of divisors as well. On the other hand, the GKZ-fan is complete, i.e., $\mathscr{M}_\mathbf{R}^- = \mathscr{M}_\mathbf{R}$, when $|\Delta|$ is strongly convex. We thus recover a result obtained by Gel'fand, Kapranov, and Zelevinskii [27] in connection with higher-dimensional hypergeometric functions. For details we refer the reader to [55], [57], [58], where "convex" should be read as "upper convex" again in the present context.

4.3. Homogeneous coordinate rings. Cox [13] recently showed a surprising result to the effect that an analog of the homogeneous coordinate ring of a projective space can be defined for a general toric variety. This result is expected to play an important role when we use toric varieties as ambient spaces as in Batyrev and Cox [7], for instance. This generalized homogeneous coordinate ring enabled Cox [13] to give a concrete description of the automorphism groups of compact toric orbifolds, generalizing that of Demazure [18] in the nonsingular case. In constructing the homogeneous coordinate ring, Cox regards a toric variety as a quotient of an open subset of an affine space, as Audin [4] and others did earlier. According to a recent result of Dolgachev and Hu [21], the quotient construction is closely related to the GKZ fan (secondary fan) in §4.2 through the geometric invariant theory of Hilbert and Mumford. In this section, we describe the gist of Cox's construction in [13] from the slightly different perspective of the equivariant holomorphic map associated to a map of fans (cf. §3.2).

Suppose that a fan Δ in $N = \mathbf{Z}^r$ satisfies the assumption $(*)$ in §3.3, and in the notation of §3.3 we define for $\widetilde{N} = \bigoplus_{\rho \in \Delta(1)} \mathbf{Z}\tilde{n}(\rho)$ a new fan $\widetilde{\Delta} := \{\tilde{\sigma} \mid \sigma \in \Delta\}$, where

$$\tilde{\sigma} := \sum_{\substack{\rho \in \Delta(1) \\ \rho \prec \sigma}} \mathbf{R}_{\geq 0}\tilde{n}(\rho) \subset \widetilde{N}_{\mathbf{R}}.$$

The scalar extension to $\mathbf{R}$ of the $\mathbf{Z}$-linear map $\pi \colon \widetilde{N} \to N$ with finite cokernel induces a surjection from $\tilde{\sigma}$ to σ. In particular, π gives rise to a map of fans from $\widetilde{\Delta}$ for $\widetilde{N}$ to Δ for N. Accordingly, we get a surjective holomorphic map $\widetilde{X} \to X$ between the corresponding toric varieties, which is equivariant with respect to the homomorphism $\widetilde{T} := \widetilde{N} \otimes_{\mathbf{Z}} \mathbf{C}^{\times} \to T$ of algebraic tori.

The set of faces of the strongly convex rational polyhedral cone

$$\sum_{\rho \in \Delta(1)} \mathbf{R}_{\geq 0}\tilde{n}(\rho) \subset \widetilde{N}_{\mathbf{R}}$$

is a fan for $\widetilde{N}$ and contains the fan $\widetilde{\Delta}$. Hence the affine space $\mathbf{C}^{\Delta(1)}$ contains $\widetilde{X}$ as a $\widetilde{T}$-stable open subset. The coordinate ring S of $\mathbf{C}^{\Delta(1)}$ coincides with the semigroup algebra over $\mathbf{C}$ of the semigroup $\sum_{\rho \in \Delta(1)} \mathbf{Z}_{\geq 0}\tilde{m}(\rho)$. For each $\rho \in \Delta(1)$, we have an element $z_\rho \in S$ corresponding to $\tilde{m}(\rho)$ so that S is the polynomial ring

$$S = \mathbf{C}[z_\rho \mid \rho \in \Delta(1)].$$

Each irreducible component of the complement of $\widetilde{X}$ in $\mathbf{C}^{\Delta(1)}$ turns out to have codimension not less than 2, so that S coincides with the ring of algebraic regular functions on $\widetilde{X}$. In view of $\mathscr{M} = \widetilde{M}/M = \sum_{\rho \in \Delta(1)} \mathbf{Z}\mu(\rho)$, Cox [13] regards S as an $\mathscr{M}$-graded ring by letting

$$\deg z_\rho := -\mu(\rho), \quad \text{for every } \rho \in \Delta(1).$$

The homogeneous part of S of degree $\mu \in \mathscr{M}$ is $S_\mu = 0$ if $\mu \notin \mathscr{M}^- := -\sum_{\rho \in \Delta(1)} \mathbf{Z}_{\geq 0}\mu(\rho)$, while S_μ for $\mu \in \mathscr{M}^-$ has

$$\left\{ \prod_{\rho \in \Delta(1)} z_\rho^{i_\rho} \,\middle|\, i_\rho \in \mathbf{Z}_{\geq 0}, \text{ for every } \rho \in \Delta(1), \text{ and } -\sum_{\rho \in \Delta(1)} i_\rho \mu(\rho) = \mu \right\}$$

as a $\mathbf{C}$-basis.

REMARK. The minus sign in $\deg z_\rho := -\mu(\rho)$ is unnatural. It is better to follow Cox [13] and define the grading of S with respect to the group of linear equivalence classes of Weil divisors as follows. As we saw in §3.4, $\mathscr{M}$ is isomorphic to the group of linear equivalence classes of Weil divisors on X by the map that sends $\mu(\rho)$ to $-v(\rho)$ for $\rho \in \Delta(1)$. The cone $\mathscr{M}^-$ corresponds under this map to the cone of linear equivalence classes of effective Weil divisors. For a sequence $(i_\rho \mid \rho \in \Delta(1)) \in$

$(\mathbf{Z}_{\geq 0})^{\Delta(1)}$ of nonnegative integers, the linear equivalence class of the effective Weil divisor $D := \sum_{\rho \in \Delta(1)} i_\rho V(\rho)$ is $[D] := \sum_{\rho \in \Delta(1)} i_\rho v(\rho)$. The degree of the monomial $z^D := \prod_{\rho \in \Delta(1)} z_\rho^{i_\rho}$ is then defined to be the linear equivalence class $[D]$.

The exact sequence $0 \to M \to \widetilde{M} \to \mathscr{M} \to 0$ of $\mathbf{Z}$-modules gives rise to an exact sequence

$$0 \leftarrow T = \operatorname{Hom}_{\mathbf{Z}}(M, \mathbf{C}^\times) \leftarrow \widetilde{T} = \operatorname{Hom}_{\mathbf{Z}}(\widetilde{M}, \mathbf{C}^\times) \leftarrow \mathscr{T} := \operatorname{Hom}_{\mathbf{Z}}(\mathscr{M}, \mathbf{C}^\times) \leftarrow 0$$

of algebraic groups. $\mathscr{T}^0 := \operatorname{Hom}_{\mathbf{Z}}(\mathscr{M}/\mathscr{M}_{\operatorname{tor}}, \mathbf{C}^\times)$ is an algebraic torus and is the connected component of the identity element of $\mathscr{T}$. The action of $\widetilde{T}$ on $\mathbf{C}^{\Delta(1)}$ induces an action of $\mathscr{T}$ on $\mathbf{C}^{\Delta(1)}$. The algebraic group $\mathscr{T}$ has $\mathscr{M}$ as the character group, and the $\mathscr{M}$-grading on the coordinate ring S induced by the action of $\mathscr{T}$ on $\mathbf{C}^{\Delta(1)}$ coincides with what we had above.

$\widetilde{T}$, hence its subgroup $\mathscr{T}$, acts algebraically on $\widetilde{X}$, and X turns out to be the categorical quotient $\widetilde{X}//\mathscr{T}$. Moreover, when Δ is simplicial, X is the geometric quotient $\widetilde{X}/\mathscr{T}$, which is much better behaved than the categorical quotient. Here is the key point of Cox's proof in [13]. If we denote by $\sigma := \pi(\tilde{\sigma}) \in \Delta$ the image of $\tilde{\sigma} \in \widetilde{\Delta}$ in $N_{\mathbf{R}}$, then the affine open subset $\widetilde{U}_{\tilde{\sigma}}$ of $\widetilde{X}$ corresponding to $\tilde{\sigma}$ has the coordinate ring $S[1/\prod_{\rho' \not\prec \sigma} z_{\rho'}]$, because of $\widetilde{M} \cap \tilde{\sigma}^\vee := \{\tilde{m} \in \widetilde{M} \mid \langle \tilde{m}, \tilde{n}(\rho) \rangle \geq 0$, for every $\rho \prec \sigma\} = \sum_{\rho \prec \sigma} \mathbf{Z}_{\geq 0}\tilde{m}(\rho) + \sum_{\rho' \not\prec \sigma} \mathbf{Z}\tilde{m}(\rho')$. Its subring of invariants with respect to the action of $\mathscr{T}$ is the semigroup ring of $M \cap \sigma^\vee := \{m \in M \mid \langle m, n(\rho) \rangle \geq 0$, for every $\rho \prec \sigma\}$, which coincides with the coordinate ring of the affine open subset U_σ of X corresponding to σ. Thus $U_\sigma = \widetilde{U}_{\tilde{\sigma}}//\mathscr{T}$.

Cox [13] associates to an $\mathscr{M}$-graded S-module a quasi-coherent $\mathscr{O}_X$-module as in the case of projective spaces due to Serre and in the case of weighted projective spaces due to Dolgachev [20].

Furthermore, when Δ is simplicial, Cox [13] succeeded in describing almost all automorphisms of the toric variety X in terms of the more manageable automorphisms of S as an $\mathscr{M}$-graded $\mathbf{C}$-algebra.

4.4. Toric Fano varieties and Calabi-Yau hypersurfaces. Fano polytopes provide an interesting class of examples of integral convex polytopes in the sense of §3.5. A convex polytope $P \subset M_{\mathbf{R}} = \mathbf{R}^r$ is said to be a *Fano polytope* if it is an r-dimensional integral (i.e., with all the vertices belonging to $M = \mathbf{Z}^r$) convex polytope containing the origin O in its interior such that for each facet (i.e., codimension-one face) P_1 of P there exists an $n_1 \in N$ satisfying

$$P_1 = \{m \in P \mid \langle m, n_1 \rangle = -1\}.$$

As in §3.5, let us consider the support function h_P and the finite complete fan Δ_P determined by P, as well as the corresponding toric projective variety $\mathbf{P}$ and the ample Cartier divisor $\mathbf{D}$. Then $\mathbf{D} = \sum_{\rho \in \Delta_P(1)} V(\rho)$ is an *anticanonical divisor*. In fact, Fano polytopes were so defined that this property holds. In general, an algebraic variety with an ample anticanonical divisor is called a Fano variety. Thus we call $\mathbf{P}$ the *toric Fano variety* associated to a Fano polytope P. It needs to be noted, however, that $\mathbf{P}$ may be singular with Gorenstein canonical singularities. The duality due to Batyrev [5], which we are about to see, was made possible thanks to this admission of mild singularities.

The *polar* of a Fano polytope P defined by

$$P^* := \{n \in N_{\mathbf{R}} \mid \langle m, n \rangle \geq -1, \text{ for every } m \in P\} \subset N_{\mathbf{R}}$$

is a convex polytope in $N_{\mathbf{R}}$ containing the origin O in its interior such that all of its vertices are in N and such that each facet P_1^* can be written as $P_1^* = \{n \in P^* \mid \langle m_1, n \rangle = -1\}$ for some $m_1 \in M$. In other words, P^* can be regarded as a Fano polytope as well, if we interchange the roles of M and N. Here is how to prove the above properties of the polar P^*. As we saw in [50, Prop. A.17], there is a duality called the *polarity* between the set of faces of P and that of P^*. In particular, the facets (resp. vertices) of P are in one-to-one correspondence with the vertices (resp. facets) of P^*. For each facet P_1 of P, the corresponding vertex $n_1 \in N$ of P^* satisfies $P_1 = \{m \in P \mid \langle m, n_1 \rangle = -1\}$, while for each vertex m_1 of P, the corresponding facet of P^* is $P_1^* = \{n \in P^* \mid \langle m_1, n \rangle = -1\}$. Moreover, the fan Δ_P turns out to be the set of strongly convex rational polyhedral cones we obtain by joining the origin O with the faces of P^*. $\{n(\rho) \mid \rho \in \Delta_P(1)\}$ is then the set of vertices of P^*. For a facet P_1^* of P^*, let σ be the r-dimensional strongly convex rational polyhedral cone we obtain by joining O with P_1^*, and let $l_\sigma \in M$ be the corresponding vertex of P. Then we have $h_P(n) = \langle l_\sigma, n \rangle$ for each $n \in \sigma$.

The Fano polytope P^* determines a fan Δ_{P^*} for M and a support function h_{P^*}, hence a toric Fano variety $\mathbf{P}^*$ with an ample Cartier divisor $\mathbf{D}^*$. The toric Fano varieties $\mathbf{P}$ and $\mathbf{P}^*$ may be said to be polars of each other.

It is an interesting nontrivial problem to classify Fano polytopes P up to coordinate transformation. Koelman [39] and Batyrev showed independently that there are 16 different two-dimensional Fano polytopes up to coordinate transformation. Batyrev and Watanabe-Watanabe showed independently that there are 18 three-dimensional Fano polytopes P with $\mathbf{P}$ nonsingular, up to coordinate transformation (cf., e.g., [50, Prop. 2.21]). According to Batyrev, there are 123 four-dimensional Fano polytopes P with $\mathbf{P}$ nonsingular, up to coordinate transformation.

The mirror symmetry of (especially three-dimensional) Calabi-Yau varieties has aroused widespread interest in connection with string theory in mathematical physics. Batyrev [5] used toric Fano varieties to systematically construct possible candidates for mirror symmetric pairs of Calabi-Yau varieties. Here is the gist of his construction.

In the toric Fano variety $\mathbf{P}$ corresponding to a Fano polytope P, consider an anticanonical hypersurface Y, that is, a closed irreducible subvariety of codimension-one linearly equivalent to the anticanonical divisor $\mathbf{D}$. Such a Y is an $(r-1)$-dimensional *Calabi-Yau variety*. Namely, a canonical divisor K_Y of Y is linearly equivalent to 0 and

$$H^j(Y, \mathcal{O}_Y) = 0 \quad \text{for every } j, \; 0 < j < \dim Y,$$

holds. If we choose Y to be sufficiently general, then Y turns out to have at worst Gorenstein singularities. Let Y^* be a general anticanonical hypersurface in the toric Fano variety $\mathbf{P}^*$ corresponding to the polar Fano polytope P^*. Then Y and Y^* are expected to form a mirror symmetric pair.

In mathematical physics, we need to modify Y and Y^* to orbifolds keeping the canonical divisors linearly equivalent to 0, and study their moduli spaces in detail. In the modification process, the GKZ-cones in §4.2 play important roles: As a modification of Y, for instance, we take a part of the inverse image (the so-called proper transform) of Y in a modification $\mathbf{P}'$ of $\mathbf{P}$. To construct such a toric Fano variety $\mathbf{P}'$, we consider all the quasi-projective subdivisions Δ' of Δ_P such that $\Delta'(1)$ is contained in $\{\mathbf{R}_{\geq 0} n \mid n \in N \cap P^*\}$. The results due to Cox [13] on the automorphism groups of toric varieties (cf. §4.3) play important roles in the process as well. For details, we refer the reader to Morrison [44], Aspinwall-Greene-Morrison [2], [3], etc.

Batyrev-Borisov [6] generalized the construction to Calabi-Yau complete intersections in toric varieties.

References

1. A. Ash, D. Mumford, M. Rapoport, and Y. Tai, *Smooth Compactification of Locally Symmetric Varieties*, Lie Groups: History, Frontiers and Applications IV, Math. Sci. Press, Brookline, Mass., 1975.
2. P. S. Aspinwall, B. R. Greene, and D. R. Morrison, *Calabi-Yau moduli space, mirror manifolds and spacetime topology change in string theory*, preprint.
3. P. S. Aspinwall, B. R. Greene, and D. R. Morrison, *The monomial-divisor mirror map* (International Mathematics Research Notes, 1993, No. 12), Duke Math. J. **72** (1993), 319–337.
4. M. Audin, *The Topology of Torus Actions on Symplectic Manifolds*, Progress in Math. vol. 93, Birkhäuser, Basel, Boston, Berlin, 1991.
5. V. V. Batyrev, *Dual polyhedra and mirror symmetry for Calabi-Yau hypersurfaces in toric varieties*, J. Algebraic Geometry **3** (1994), 493–535.
6. V. V. Batyrev and L. A. Borisov, *Dual cones and mirror symmetry for generalized Calabi-Yau manifolds*, preprint.
7. V. V. Batyrev and D. Cox, *On the Hodge structure of projective hypersurfaces in toric varieties*, Duke Math. J. **75** (1994), 293–338.
8. A. A. Beilinson, J. Bernstein, and P. Deligne, *Faisceaux pervers*, in Analyse et Topologie sur les Espaces Singuliers (I), Astérisque, vol. 100, Soc. Math. France, 1982.
9. M. Brion, *Points entiers dans les polyèdres convexes*, Ann. Sci. École Norm. Sup. (4) **21** (1988), 653–663.
10. M. Brion, *Polyèdres et réseaux*, Enseign. Math. (2) **38** (1992), 71–88.
11. M. Brion, *Points entiers dans les polytopes convexes*, Séminaire Bourbaki, exp. 780, March, 1994.
12. L. Caporaso, *A compactification of the universal Picard variety over the moduli space of stable curves*, J. Amer. Math. Soc. **7** (1994), 589–660.
13. D. Cox, *The homogeneous coordinate ring of a toric variety*, J. Algebraic Geometry (to appear).
14. R. Dabrowski, *On normality of the closure of a generic torus orbit in G/P*, preprint.
15. V. I. Danilov, *The geometry of toric varieties*, Russian Math. Surveys **33** (1978), 97–154; Uspehi Mat. Nauk **33** (1978), 85–134.
16. V. I. Danilov, *The birational geometry of toric 3-folds*, Math. USSR-Izv. **21** (1983), 269–280; Izv. Akad. Nauk SSSR, Ser. Mat. **46** (1982), 971–982.
17. V. I. Danilov, *De Rham complex on toroidal variety*, in Algebraic Geometry (S. Bloch, I. Dolgachev, and W. Fulton, eds.), Lecture Notes in Math., vol. 1479, Springer-Verlag, Berlin, Heidelberg, New York, 1991, pp. 26–38.
18. M. Demazure, *Sous-groupes algébriques de rang maximum du groupe de Cremona*, Ann. Sci. École Norm. Sup. (4) **3** (1970), 507–588.
19. J. Denef and F. Loeser, *Weights of exponential sums, intersection cohomology, and Newton polyhedra*, Invent. Math. **106** (1991), 275–294.
20. I. Dolgachev, *Weighted projective varieties*, in Group Actions and Vector Fields, Proceedings, Vancouver 1981 (J. B. Carrell, ed.), Lecture Notes in Math., vol. 956, Springer-Verlag, Berlin, Heidelberg, New York, 1982, pp. 34–71.
21. I. Dolgachev and Y. Hu, *Variation of geometric invariant theory quotients*, preprint.
22. G. Ewald, *Algebraic geometry and convexity*, in Handbook of Convex Geometry (P. M. Gruber and J. M. Wills, eds.), Vols. A and B, North-Holland, Amsterdam, 1993, Chapter 2.6, pp. 603–626.
23. G. Ewald, *Combinatorial Convexity and Algebraic Geometry*, Graduate Texts in Math., Springer-Verlag, Berlin (to appear).
24. K.-H. Fieseler, *Rational intersection cohomology of projective toric varieties*, J. Reine Angew. Math. **413** (1991), 88–98.
25. W. Fulton, *Introduction to Toric Varieties*, Ann. of Math. Studies, vol. 131, Princeton Univ. Press, 1993.

26. W. Fulton and B. Sturmfels, *Intersection theory on toric varieties*, preprint.
27. I. M. Gel′fand, A. V. Zelevinskii, and M. M. Kapranov, *Discriminants of polynomials in several variables and triangulations of Newton polyhedra*, Leningrad Math. J. **2** (1991), no. 3, 449–505; Algebra i Analiz. **2** (1990), no. 3, 1–62.
28. M. Goresky and R. MacPherson, *Intersection homology theory*, Topology **19** (1980), 135–162.
29. M. Goresky and R. MacPherson, *Intersection homology* II, Invent. Math. **72** (1983), 77–129.
30. T. Hibi, *Combinatorics on simplicial complexes and convex polytopes* (in Japanese), Sugaku, Math. Soc. Japan **44** (1992), 147–160; English translation: Sugaku Expositions, Amer. Math. Soc., Providence, RI (to appear).
31. M.-N. Ishida, *Torus embeddings and de Rham complexes*, in Commutative Algebra and Combinatorics (M. Nagata and H. Matsumura, eds.), Advanced Studies in Pure Math., vol. 11, Kinokuniya, Tokyo and North-Holland, Amsterdam, New York, Oxford, 1987, pp. 111–145.
32. M.-N. Ishida, *Polyhedral Laurent series and Brion's equalities*, Internat. J. Math. **1** (1990), 251–265.
33. M.-N. Ishida, *The duality of cusp singularities*, Math. Ann. **294** (1992), 81–97.
34. M.-N. Ishida, *Torus embeddings and algebraic intersection complexes*, preprint.
35. M.-N. Ishida, *Torus embeddings and algebraic intersection complexes*, II, preprint.
36. T. Kajiwara, *Logarithmic compactifications of the generalized Jacobian variety*, J. Fac. Sci. Univ. Tokyo, Sec. IA **40** (1993), 473–502.
37. K. Kato, *Logarithmic structures of Fontaine-Illusie*, in Algebraic Analysis, Geometry, and Number Theory, Proceedings of the JAMI Inaugural Conference (J.-I. Igusa, ed.), Supplement to Amer. J. Math., The Johns Hopkins Univ. Press, 1989, pp. 191–224.
38. G. Kempf, F. Knudsen, D. Mumford, and B. Saint-Donat, *Toroidal Embeddings* I, Lecture Notes in Math., vol. 339, Springer-Verlag, Berlin, Heidelberg, New York, 1973.
39. R. J. Koelman, *The number of moduli of families of curves on toric surfaces*, thesis, Univ. Nijmegen, 1991.
40. T. Mabuchi, *Einstein-Kähler forms, Futaki invariants and convex geometry on toric Fano varieties*, Osaka J. Math. **24** (1987), 705–737.
41. P. McMullen, *On simple polytopes*, Invent. Math. **113** (1993), 419–444.
42. K. Miyake and T. Oda, *Almost homogeneous algebraic varieties under algebraic torus actions*, in Manifolds—Tokyo, 1973 (A. Hattori, ed.), Univ. of Tokyo Press, 1975, pp. 373–381.
43. R. Morelli, *The birational geometry of toric varieties*, preprint.
44. D. R. Morrison, *Mirror symmetry and rational curves on quintic threefolds: A guide for mathematicians*, J. Amer. Math. Soc. **6** (1993), 223–247.
45. Y. Nakagawa, *Einstein-Kähler toric Fano fourfolds*, Tohoku Math. J. **45** (1993), 297–310.
46. Y. Nakagawa, *Classification of Einstein-Kähler toric Fano fourfolds*, Tohoku Math. J. **46** (1994), 125–133.
47. T. Oda, *Lectures on Torus Embeddings and Applications (Based on Joint Work with Katsuya Miyake)*, Tata Inst. of Fund. Research, vol. 58, Springer-Verlag, Berlin, Heidelberg, New York, 1978.
48. T. Oda, *The geometry of convex bodies and algebraic geometry*, Sugaku, Math. Soc. Japan **33** (1981), 120–133. (Japanese)
49. T. Oda, *Convex Bodies and Algebraic Geometry*, Kinokuniya, Tokyo, 1985. (Japanese)
50. T. Oda, *Convex Bodies and Algebraic Geometry—An Introduction to the Theory of Toric Varieties*, Ergebnisse der Math. (3), vol. 15, Springer-Verlag, Berlin, Heidelberg, New York, London, Paris, Tokyo, 1988.
51. T. Oda, *Simple convex polytopes and the strong Lefschetz theorem*, J. Pure Appl. Algebra **71** (1991), 265–286.
52. T. Oda, *Geometry of toric varieties*, in Proc. of the Hyderabad Conf. on Algebraic Groups (S. Ramanan, ed.), Manoj Prakashan, Madras, 1991, pp. 407–440.
53. T. Oda, *The algebraic de Rham theorem for toric varieties*, Tohoku Math. J. **45** (1993), 231–247.
54. T. Oda, *Convex bodies and algebraic geometry*, Surikagaku, March 1994, pp. 36–40. (Japanese)
55. T. Oda and H. S. Park, *Linear Gale transforms and Gelfand-Kapranov-Zelevinskij decompositions*, Tohoku Math. J. **43** (1991), 375–399.
56. T. Oda and C. S. Seshadri, *Compactifications of the generalized Jacobian variety*, Trans. Amer. Math. Soc. **253** (1979), 1–90.
57. H. S. Park, *Algebraic cycles on toric varieties*, thesis, Tohoku Univ., March, 1992.
58. H. S. Park, *The Chow rings and GKZ-decompositions for $\mathbf{Q}$-factorial toric varieties*, Tohoku Math. J. **45** (1993), 109–145.
59. S. Sardo Infirri, *Lefschetz fixed-point theorem and lattice points in convex polytopes*, preprint.
60. I. Satake, *On the arithmetic of tube domains*, Bull. Amer. Math. Soc. **79** (1973), 1076–1094.

61. R. Stanley, *Generalized h-vectors, intersection cohomology of toric varieties, and related results*, in Commutative Algebra and Combinatorics (M. Nagata and H. Matsumura, eds.), Advanced Studies in Pure Math., vol. 11, Kinokuniya, Tokyo and North-Holland, Amsterdam, New York, Oxford, 1987, pp. 187–213.
62. R. Stanley, *Subdivisions and local h-vectors*, J. Amer. Math. Soc. **5** (1992), 805–851.
63. H. Sumihiro, *Equivariant completion*, I, II, J. Math. Kyoto Univ. **14** (1974), 1–28; ibid. **15** (1975), 573–605.
64. J. Włodarczyk, *Decompositions of birational toric maps in blow-ups and blow-downs. A proof of the weak Oda conjecture*, preprint.

MATHEMATICAL INSTITUTE, FACULTY OF SCIENCE, TOHOKU UNIVERSITY, SENDAI 980-77, JAPAN
E-mail address: odatadao@math.tohoku.ac.jp

Translated by TADAO ODA

Other Titles in This Series

(Continued from the front of this publication)

133 R. R. Suncheleev et al., Thirteen Papers in Analysis
132 I. G. Dmitriev et al., Thirteen Papers in Algebra
131 V. A. Zmorovich et al., Ten Papers in Analysis
130 M. M. Lavrent′ev, K. G. Reznitskaya, and V. G. Yakhno, One-dimensional Inverse Problems of Mathematical Physics
129 S. Ya. Khavinson, Two Papers on Extremal Problems in Complex Analysis
128 I. K. Zhuk et al., Thirteen Papers in Algebra and Number Theory
127 P. L. Shabalin et al., Eleven Papers in Analysis
126 S. A. Akhmedov et al., Eleven Papers on Differential Equations
125 D. V. Anosov et al., Seven Papers in Applied Mathematics
124 B. P. Allakhverdiev et al., Fifteen Papers on Functional Analysis
123 V. G. Maz′ya et al., Elliptic Boundary Value Problems
122 N. U. Arakelyan et al., Ten Papers on Complex Analysis
121 V. D. Mazurov, Yu. I. Merzlyakov, and V. A. Churkin, Editors, The Kourovka Notebook: Unsolved Problems in Group Theory
120 M. G. Kreĭn and V. A. Jakubovič, Four Papers on Ordinary Differential Equations
119 V. A. Dem′janenko et al., Twelve Papers in Algebra
118 Ju. V. Egorov et al., Sixteen Papers on Differential Equations
117 S. V. Bočkarev et al., Eight Lectures Delivered at the International Congress of Mathematicians in Helsinki, 1978
116 A. G. Kušnirenko, A. B. Katok, and V. M. Alekseev, Three Papers on Dynamical Systems
115 I. S. Belov et al., Twelve Papers in Analysis
114 M. Š. Birman and M. Z. Solomjak, Quantitative Analysis in Sobolev Imbedding Theorems and Applications to Spectral Theory
113 A. F. Lavrik et al., Twelve Papers in Logic and Algebra
112 D. A. Gudkov and G. A. Utkin, Nine Papers on Hilbert's 16th Problem
111 V. M. Adamjan et al., Nine Papers on Analysis
110 M. S. Budjanu et al., Nine Papers on Analysis
109 D. V. Anosov et al., Twenty Lectures Delivered at the International Congress of Mathematicians in Vancouver, 1974
108 Ja. L. Geronimus and Gábor Szegő, Two Papers on Special Functions
107 A. P. Mišina and L. A. Skornjakov, Abelian Groups and Modules
106 M. Ja. Antonovskiĭ, V. G. Boltjanskiĭ, and T. A. Sarymsakov, Topological Semifields and Their Applications to General Topology
105 R. A. Aleksandrjan et al., Partial Differential Equations, Proceedings of a Symposium Dedicated to Academician S. L. Sobolev
104 L. V. Ahlfors et al., Some Problems on Mathematics and Mechanics, On the Occasion of the Seventieth Birthday of Academician M. A. Lavrent′ev
103 M. S. Brodskiĭ et al., Nine Papers in Analysis
102 M. S. Budjanu et al., Ten Papers in Analysis
101 B. M. Levitan, V. A. Marčenko, and B. L. Roždestvenskiĭ, Six Papers in Analysis
100 G. S. Ceĭtin et al., Fourteen Papers on Logic, Geometry, Topology and Algebra
99 G. S. Ceĭtin et al., Five Papers on Logic and Foundations
98 G. S. Ceĭtin et al., Five Papers on Logic and Foundations
97 B. M. Budak et al., Eleven Papers on Logic, Algebra, Analysis and Topology
96 N. D. Filippov et al., Ten Papers on Algebra and Functional Analysis
95 V. M. Adamjan et al., Eleven Papers in Analysis
94 V. A. Baranskiĭ et al., Sixteen Papers on Logic and Algebra

(See the AMS catalog for earlier titles)